Lecture Notes in Mechanical Engineering

Lecture Notes in Mechanical Engineering (LNME) publishes the latest developments in Mechanical Engineering - quickly, informally and with high quality. Original research reported in proceedings and post-proceedings represents the core of LNME. Volumes published in LNME embrace all aspects, subfields and new challenges of mechanical engineering. Topics in the series include:

- Engineering Design
- Machinery and Machine Elements
- Mechanical Structures and Stress Analysis
- Automotive Engineering
- Engine Technology
- Aerospace Technology and Astronautics
- Nanotechnology and Microengineering
- Control, Robotics, Mechatronics
- MEMS
- Theoretical and Applied Mechanics
- Dynamical Systems, Control
- Fluid Mechanics
- Engineering Thermodynamics, Heat and Mass Transfer
- Manufacturing
- Precision Engineering, Instrumentation, Measurement
- Materials Engineering
- Tribology and Surface Technology

To submit a proposal or request further information, please contact the Springer Editor in your country:

China: Li Shen at li.shen@springer.com
India: Dr. Akash Chakraborty at akash.chakraborty@springernature.com
Rest of Asia, Australia, New Zealand: Swati Meherishi at swati.meherishi@springer.com
All other countries: Dr. Leontina Di Cecco at Leontina.dicecco@springer.com

To submit a proposal for a monograph, please check our Springer Tracts in Mechanical Engineering at http://www.springer.com/series/11693 or contact Leontina.dicecco@springer.com

Indexed by SCOPUS. The books of the series are submitted for indexing to Web of Science.

More information about this series at http://www.springer.com/series/11236

Maciej Majewski · Wojciech Kacalak
Editors

Innovations Induced by Research in Technical Systems

Springer

Editors
Maciej Majewski
Faculty of Mechanical Engineering
Koszalin University of Technology
Koszalin, Poland

Wojciech Kacalak
Faculty of Mechanical Engineering
Koszalin University of Technology
Koszalin, Poland

ISSN 2195-4356 ISSN 2195-4364 (electronic)
Lecture Notes in Mechanical Engineering
ISBN 978-3-030-37565-2 ISBN 978-3-030-37566-9 (eBook)
https://doi.org/10.1007/978-3-030-37566-9

This Springer imprint is published by the registered company Springer Nature Switzerland AG
The registered company address is: Gewerbestrasse 11, 6330 Cham, Switzerland

Preface

This volume of Lecture Notes in Mechanical Engineering contains selected papers from the International online Conference on Innovations Induced by Research in Technical Systems - IIRTS 2019, held on 15–18 October 2019.

The conference was organized by the Department of Technical and Informatics Systems Engineering - Faculty of Mechanical Engineering, Koszalin University of Technology, Poland, under the scientific auspices of the Committee on Machine Building of the Polish Academy of Sciences (PAN).

The Conference IIRTS serves as a platform for researchers, academicians and professionals to present, discuss and promote innovative solutions and approaches to problems created as a result of their research related to scientific and engineering inventions in the broad field of technical systems. The main conference topics were:

- Modelling of decision-making processes in machine design and manufacturing
- Innovative technological processes in machine construction
- Processes of innovation design and implementation
- Economic aspects of structural and technological innovations
- Knowledge management in industrial organizations
- Automation of production processes
- Algorithmization and digitalization of problems in machine design and operation
- Monitoring of technological processes
- Quality engineering
- Human–machine interface design and information systems
- Human–machine interaction and teaming
- Industrial applications of artificial intelligence and expert systems.

The organizers received contributions from many countries around the world. After a thorough peer review process, the committee accepted 17 papers for the conference proceedings prepared by 30 authors.

We would like to thank the members of the International Scientific Committee for their contributions and the reviews. We would also like to thank the authors for their interesting papers and presentations.

We would like to thank the publisher, Springer, for cooperation in publishing the proceedings in the prestigious series of Lecture Notes in Mechanical Engineering. Great support and collaboration from Springer, in particular from LNME Editor, Dr. Leontina Di Cecco, are greatly appreciated.

Wojciech Kacalak
Maciej Majewski

Organization

International Scientific Committee Chairs

Wojciech Kacalak	Koszalin University of Technology, Poland
Maciej Majewski	Koszalin University of Technology, Poland

International Scientific Committee Members

Chamil Abeykoon	The University of Manchester, UK
Andrzej Ambroziak	Wrocław University of Science and Technology, Poland
Charis Apostolopoulos	University of Patras, Greece
Jan Awrejcewicz	Lodz University of Technology, Poland
Jerzy Bajkowski	Warsaw University of Technology, Poland
Marek Balazinski	Polytechnique Montréal, Canada
Sergio Baragetti	University of Bergamo, Italy
Adam Barylski	Gdańsk University of Technology, Poland
Stefan Berczyński	West Pomeranian University of Technology, Poland
Tadeusz Bohdal	Koszalin University of Technology, Poland
Zbigniew Budniak	Koszalin University of Technology, Poland
Xun Chen	Liverpool John Moores University, UK
Olaf Ciszak	Poznan University of Technology, Poland
Jan Duda	Cracow University of Technology, Poland
Stanisław Duer	Koszalin University of Technology, Poland
Isaac Elishakoff	Florida Atlantic University, USA
Józef Gawlik	Cracow University of Technology, Poland
Zbigniew Gronostajski	Wrocław University Science of Technology, Poland
Adam Hamrol	Poznan University of Technology, Poland
Jean-Yves Hascoët	École Centrale de Nantes, France
Tadeusz Hryniewicz	Koszalin University of Technology, Poland

Jan Kiciński	Institute of Fluid-Flow Machinery, Poland
Janusz Kowal	AGH University of Science and Technology, Poland
Dražan Kozak	University of Osijek, Croatia
Waldemar Kuczyński	Koszalin University of Technology, Poland
Leon Kukiełka	Koszalin University of Technology, Poland
Dariusz Lipiński	Koszalin University of Technology, Poland
Czesław Łukianowicz	Koszalin University of Technology, Poland
Krzysztof Marchelek	West Pomeranian University of Technology, Poland
Tadeusz Markowski	Rzeszów University of Technology, Poland
Thomas Mathia	École Centrale de Lyon, France
Adam Mazurkiewicz	Institute for Sustainable Technologies, Poland
Slobodan Mitrovic	University of Kragujevac, Serbia
Andrew Y. C. Nee	National University of Singapore, Singapore
Mirosław Pajor	West Pomeranian University of Technology, Poland
François Pérès	Institut National Polytechnique de Toulouse-ENIT, France
Roman Staniek	Poznan University of Technology, Poland
Ion Stiharu	Concordia University, Canada
Marian Szczerek	Institute for Sustainable Technologies, Poland
Marian Wiercigroch	University of Aberdeen, UK
Jan Żurek	Poznan University of Technology, Poland

Organizing Committee

Chair

Maciej Majewski	Koszalin University of Technology, Poland

Member

Filip Szafraniec	Koszalin University of Technology, Poland

Contents

Quality Assessment Using Antipatterns in Machine Building and Operations

Andrzej Tuchołka(✉), Maciej Majewski, Wojciech Kacalak, and Zbigniew Budniak

Faculty of Mechanical Engineering, Koszalin University of Technology, Raclawicka 15-17, 75-620 Koszalin, Poland
andrzej.tucholka@gmail.com

Abstract. Authors propose a method to classify the antipatterns, allowing their further definition for manual and automatic usage. We apply antipatterns in quality assessment as a negative quality reference. This enables counting and measuring similarity to predictable errors (antipatterns), automatic identification of erroneous structures found in mechanical designs, or anomalies in machine operations. We present approaches in which innovative use of antipatterns allows technical employees to improve their existing processes. In essence, we have found that antipatterns are useful in design, monitoring, and quality assurance of machine elements during their design, operations, and maintenance.

Keywords: Antipatterns · Quality assessment · Neural networks

1 Introduction

Several successful approaches to quality assessment of mechanical constructions can be found in the history. They are mainly based on the verification of manual and automated physical models, computer based physical simulations, and a widespread use of expert knowledge. Our method introduces and focuses on antipatterns (i.e. incorrect repeatable solutions for a given functional problem) to identify errors in the input data describing machine element(s). Our approach is meant to support existing methods, by providing a relatively quick and low cost method of including the shared knowledge about possible errors in quality assessments.

Common quality models are based on the perfect behavior, leaving little room for innovative solutions to given problems. We aim to open the set of possibilities by judging them with regards to their fitness for the expected function, rather than proximity to the idealized solution. This approach cannot be considered as an exhaustive analysis, but provides an opportunity to increase the quality of design, operations, by allowing for automated detection of errors through similarity comparison with known and incorrect behavior or solutions.

Analysis of antipatterns has been first conceptualized and identified in computer science [1] and since the concept has been slowly adopted by other science disciplines. We provide concrete examples of applications of antipatterns in machine construction and operations [3, 4, 9].

M. Majewski and W. Kacalak (Eds.): IIRTS 2019, LNME, pp. 1–12, 2020.
https://doi.org/10.1007/978-3-030-37566-9_1

2 Antipattern Identification

Symbolic representation of erroneous data patterns, for the purpose of detecting similarity, can be described with regards to the origin of the error (incorrect feature value, incorrect relation between feature values, incorrect structure with feature values) and the phase in its lifetime (design, production, operations, recall).

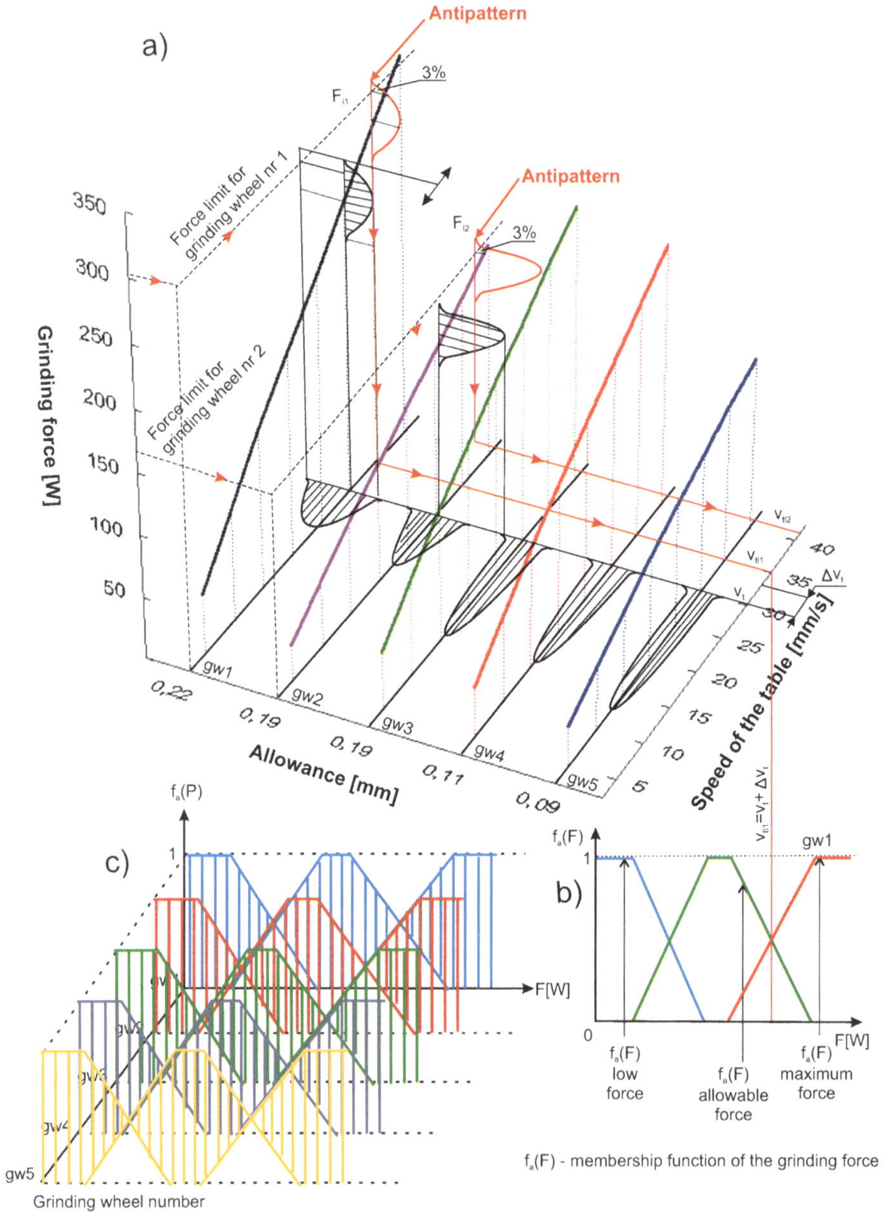

Fig. 1. Algorithm for assessing the technological safety of commands based on the real process.

One of the approaches we have analyzed is the safety analysis of commands concerning grinding force, machining allowance and the possibility of increasing the table speed. Taking into account the features of the real grinding process an algorithm was created for assessing the technological safety of commands provided to the machine in the context of its current operational parameters of the real technological process. A model of grinding force was developed as a function of feed speed and machining allowance.

In (Fig. 1), the lines represent dependence of the force on the grinding process parameters for particular grinding wheels. Basing on the specified criteria, there is the grinding force limit determined for each grinding wheel. Basing on the grinding force limit, there is the table speed limit assigned. According to the operator's command, if the increase of the speed makes a speed of the table smaller than the smallest speed determined from the force limit for all the grinding wheels, then the command is safe to be executed. Such relatively no-trivial algorithm allows for a continuous monitoring and improvement of matching the grinding force with the speed by observing occurrence of antipatterns in the operational configuration of the machine.

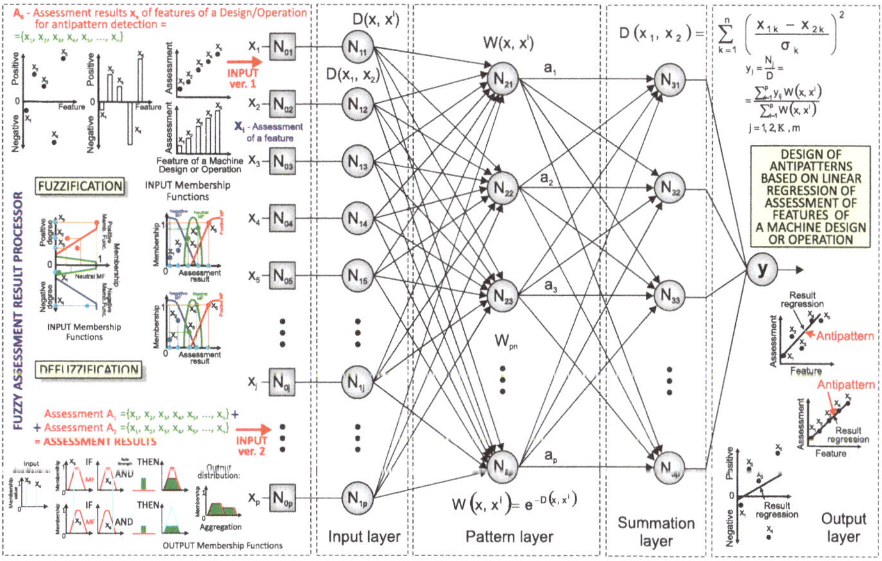

Fig. 2. Regression neural network design for antipattern identification.

Using neural networks it is also possible to mimic the expert behavior, where the regression algorithms embedded in the neural network, allow to automatically identify antipatterns (Fig. 2). Following this approach the incorrectness of the input data is filtered and normalized to allow for qualitative grading of construction features or machine's operational states. Such antipattern identification algorithms can be applied universally (in terms of detection and description of the antipatterns) but require manual analysis and network configuration to adapt the algorithms to the data patterns found in the input. Additionally it is required to design and define the correctness criteria for processed technological values, such that they can be reflected in the network design. Provided a consistent data normalization process these networks can be applied for meaningful interpretation of mechanical construction designs, or analysis of operational signals.

Further challenges in identifying antipatterns lie in normalization challenges, where the data patterns have to be clearly identifiable, yet vary such that the training process doesn't result in over fitting of the network parameters to few training samples. This is commonly addressed using data blurring and gradient functions. In case of structured data with implicit relations, it is necessary to map possible relations occurring among all of the analyzed features (Fig. 3).

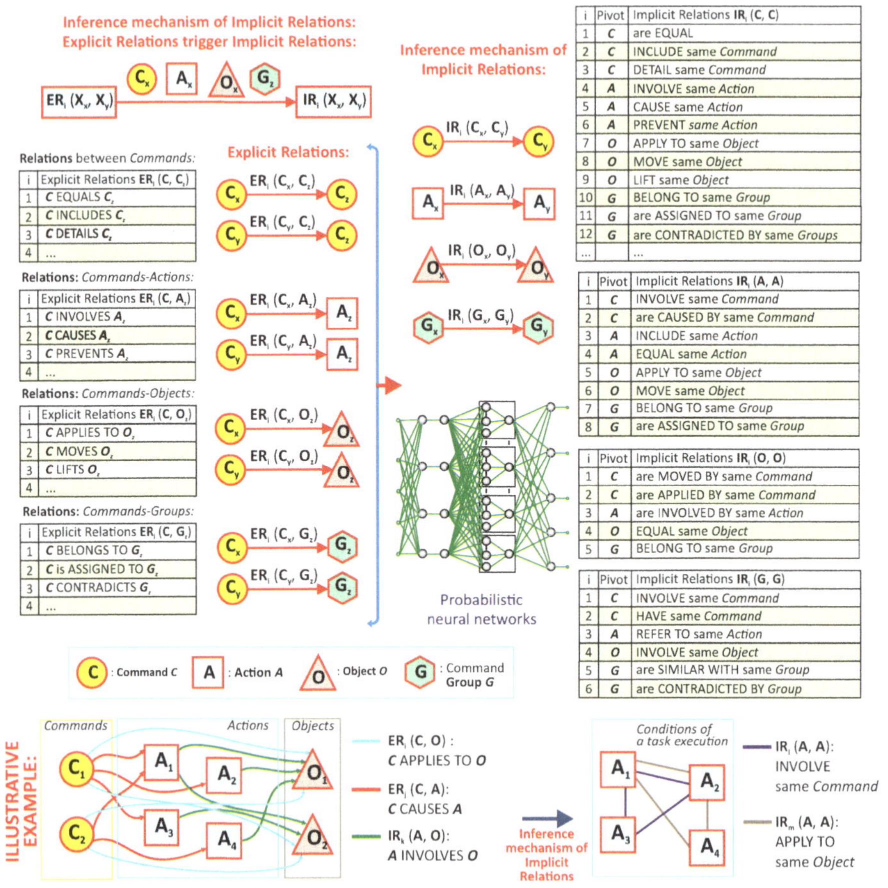

Fig. 3. Model for relations inference in machine operations.

We have found lexical decomposition and analysis of the element's function to be a required component enabling inclusion of expert knowledge (embodied in antipatterns) for use by automated methods for antipattern identification, definition, and detecting similarity. Here (Fig. 3) the key lexical components relate to machine operations (command, action, object), as a result of the need to reflect on the operational nature of the data.

3 Antipattern Classes

We use antipatterns as a negative quality reference, which means that the similarity to an antipattern represents a reduction in efficiency and fitness for the purpose and intended function. This approach enables incorporation of experience based knowledge, allowing to incorporate both: known errors, and ones that are automatically detected. Such automation of the expert knowledge is important for the computers to better react to the problems we face each time some machine element is designed, produced, operated, or recalled. Apart from the open-ended ability to continuously identify antipatterns, there is a large set of expert knowledge (regarding errors in that phase) that can be enclosed in the antipattern model for further automatic processing.

We started with antipatterns in constructions and saw a wider application of our research in at least other parts of machine element's life cycle. These antipatterns follow a similar abstraction model and can be found in: description of the design, measurements during/after production, physical properties observed during machine's operations, and during the validation and identification of element's recalled parts (Table 1).

Table 1. Antipattern classes with examples.

Phase	Values	Relations	Structures
Machine design	Wrong roughness of the surface	Lack of ribbing on the corpse of a gearbox	Rigging placement on the gearbox
Production of the machine	Wrong material used for the case	Misaligned axis of shafts in a gearbox	Misplaced weld on the gearbox corpse
Operations of the machine	Vibrations level of the gearbox case	Worn out threads of the gearbox's shafts	Mismatched ball bearing
Recall of the machine	Highly polluted Material	Pierced corpse of the gearbox	Missing ball in a ball bearing

4 Antipatterns in Machine Construction

In each phase of the machine's lifetime it is possible to identify and meaningfully apply the method of detecting similarities to known antipatterns [3, 4, 9] to increase the overall quality of the process. As in other phases, the antipatterns can be identified in designs of incorrect feature values (Fig. 4), relations between feature values (Fig. 5), and structural errors (Fig. 6).

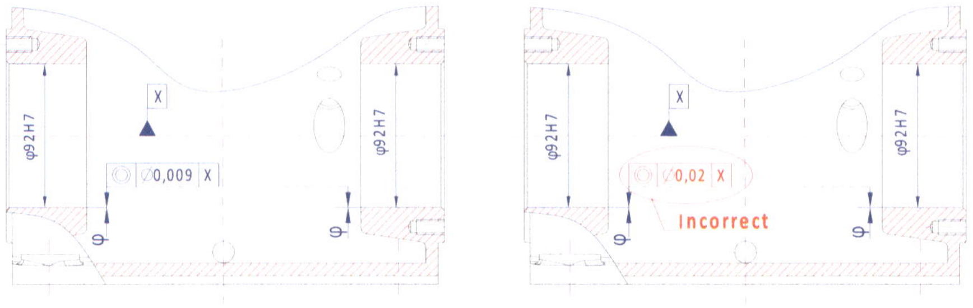

Fig. 4. Value based antipattern - tolerance of surface precision in ball bearing holes.

Feature value based antipatterns, are ones that are easiest to identify and correct. The incorrectness of the design arises from assigning a wrong value to a feature of an element. Because such wrong value changes the properties of the element, and such a change has a negative effect on the function of the element, we can consider it an antipattern. In the example above (Fig. 4) the feature of an element in question is the precision of the surface in the ball bearing hole.

More complex antipatterns are found when looking at the relations in the construction and between many feature values. The incorrectness of these antipatterns arise from negative properties observed after combining multiple elements. These sometimes implicit relations between parts of the construction, define the final fitness of the design for the intended function. In the example below (Fig. 5) the welding should be placed at the end of the corpse, but depending on its position with regards to elements and this placement can be correct or an antipattern. This positioning problem is caused by the difference in how the forces will act on the welding and how well the weld will hold the structure together.

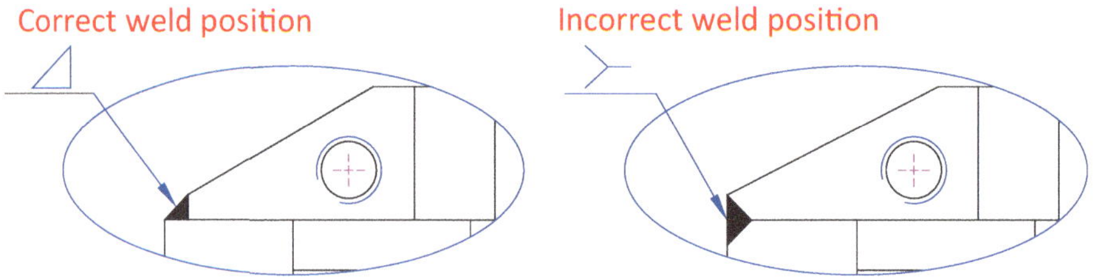

Fig. 5. Relation based antipattern - incorrect position of the weld on the gearbox's corpse.

It is important to note that the design antipatterns found in feature values can only be explicit (i.e. the incorrect feature value that has to be listed or missing). Once we start taking into consideration the structural relations and structure of the element, the limitations of the technical drawings (traditionally used to represent mechanical designs) create a set of implicit relations that can be noticed only when applying expert knowledge or computer simulation [7]. This limitation is directly related to the incompleteness and lack of contextual information (i.e. fitness for assumed function).

In case of the weld positioning (Fig. 5) the drawing is missing the information on kinetic forces that will be affecting the weld. Having these represented in the drawing, the antipattern could be reduced to detection of limited strength of the welding joint. In the structural example below (Fig. 6) the designer has defined the slope of the gearbox's floor simply by increasing the height of the part of the floor opposite to the drain hole. This defined an implicit feature value – the angle of the slope of the floor.

The structural antipatterns (e.g. incorrect slope of the gearbox's floor) are the third class of antipatterns we have analyzed. Here, the function of the element that is failing is the ability to drain all of oil from the case, and as much as a similar antipattern could be represented with a simple angle of the bottom surface, in this particular case, there is no such thing as the bottom surface, and a proper slope is a result of a structural correlation of the height feature of parts of the base of the gearbox.

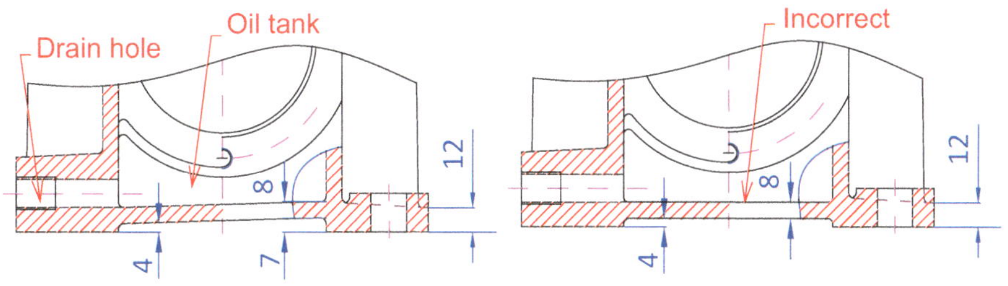

Fig. 6. Structural antipattern - incorrect slope of the gearbox's floor.

Technical drawings, with all their limitations (e.g. lack of functional definition, implicit feature values) provide a set of precise and direct measurements of features that describe the machine element [2, 6]. In later stages of machine's lifetime, access to such information is not guaranteed, but due to the universal nature of antipatterns, even given a data set with a description of indirect features, it is possible to detect in it data patterns and reason upon them.

The benefits of using antipatterns to increase machine design's quality comes mainly from embodiment of the common design knowledge into antipatterns and automating their detection in designed elements [8, 10]. Such a common library could decrease amount of faulty machine designs with relatively low cost.

In the production phase, apart from the problems arising during the production process itself, we have found that value antipatterns can be used for normalization of feature values itself, namely to define the boundaries of the acceptable solution's space using antipatterns. On Fig. 7, we present how each of the parameters of the process identifying configuration for the grinding wheel (e.g. treatment precision) is normalized to a 0…1 value space, with bounds set with antipatterns representing max and min feature values. A set of such input properties is then processed by a numerical algorithm to identify correct configuration of features to allow for the correct grinding wheel to be chosen in a fully automated manner.

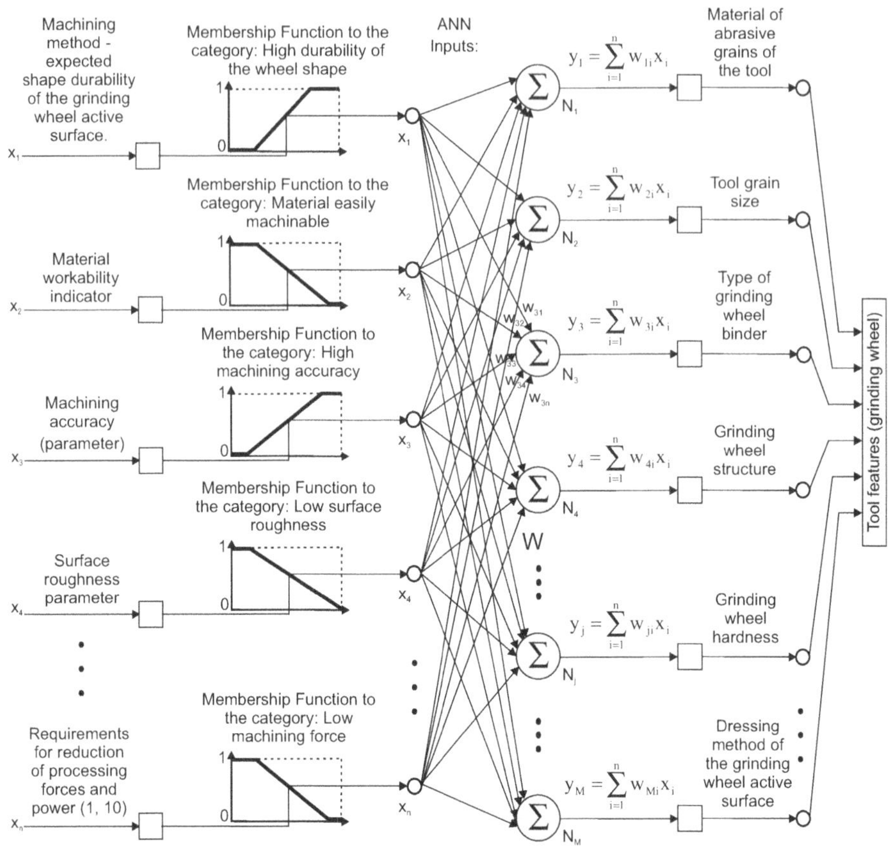

Fig. 7. Antipatterns used for normalization of the feature requirements.

5 Antipatterns in Machine Operations

We have also identified several ways in which antipatterns can be applied to detect the correlation between the description of the machine operations and the quality of its work. In case of machine operations antipatterns can be identified in most of the data features it generates (both direct and indirect). The mapping between the antipatterns found in the signal and the fitness of the machine for a given function, can again be observed in the operational signals and the operational configuration (similar to the production phase).

As demonstrated, antipatterns can be observed in most of normalized datasets describing mechanical constructions. Operational aspects of modern mechanical constructions often depend on the complex configuration provided by their operators. We have analyzed these external (to the machine) datasets and were also able to identify antipatterns with significant qualitative impact on the machine operations. On Fig. 8, we present one of the approaches to analysis of the machine (i.e. mobile crane) operations performed in task groups. This approach [5] (supported with neural networks), enables inference on the implicit relations found between input data samples.

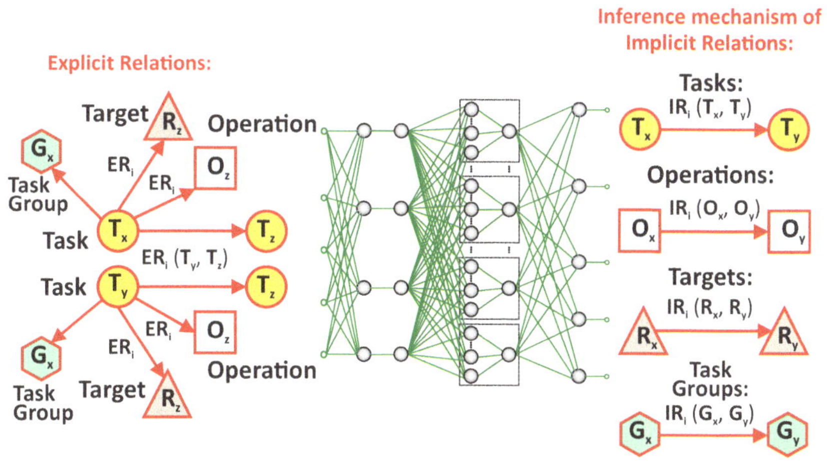

Fig. 8. An approach to identify implicit antipattern relations in the operational plan.

Antipatterns can be used as a qualitative reference allowing for automatic detection of incorrect execution plans. Following this approach we can use antipatterns to validate operational plans of the machines against a pre-defined library of antipatterns, removing the occurrence of common errors.

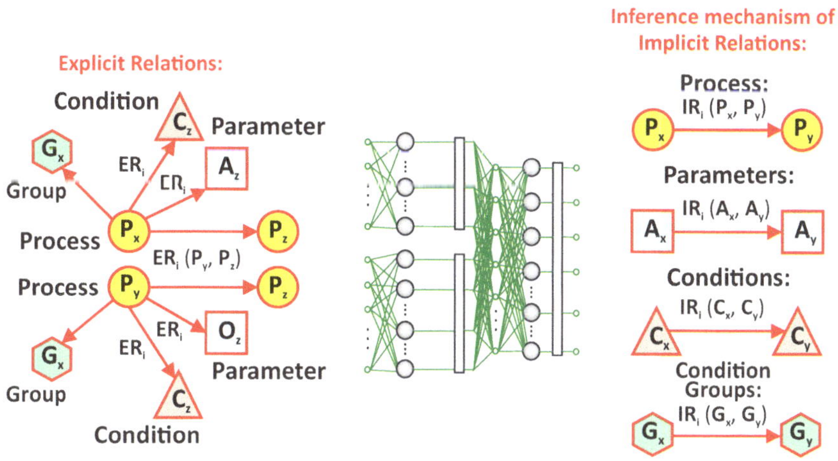

Fig. 9. An approach to identify implicit antipattern relations in operational configuration.

The abstract nature of antipatterns, allows for easy adaptations to the specifics of the context. A sibling example (Fig. 9) presents the inference model enabling detection of implicit antipatterns in crane's operational configuration. It requires updated numerical models to account for changed logic and restructured domain language, but at the essence it's applied in the same way - to detect known errors in structured data – antipatterns.

Identification, definition, and detection of antipatterns is also useful when monitoring the execution of the task plan and associated operations. Figure 10 presents a model of neural networks allowing for automatic validation of operational configuration. In this case, the algorithm processes the input data - conditions and parameters of the operation performed by the machine. These input conditions, filtered by their relevance, are mapped to the neural network trained with pre-defined antipattern values. Detecting similarity to antipatterns, allows for a quick interpretation of machine's configuration, detection of known errors, and preventing failures.

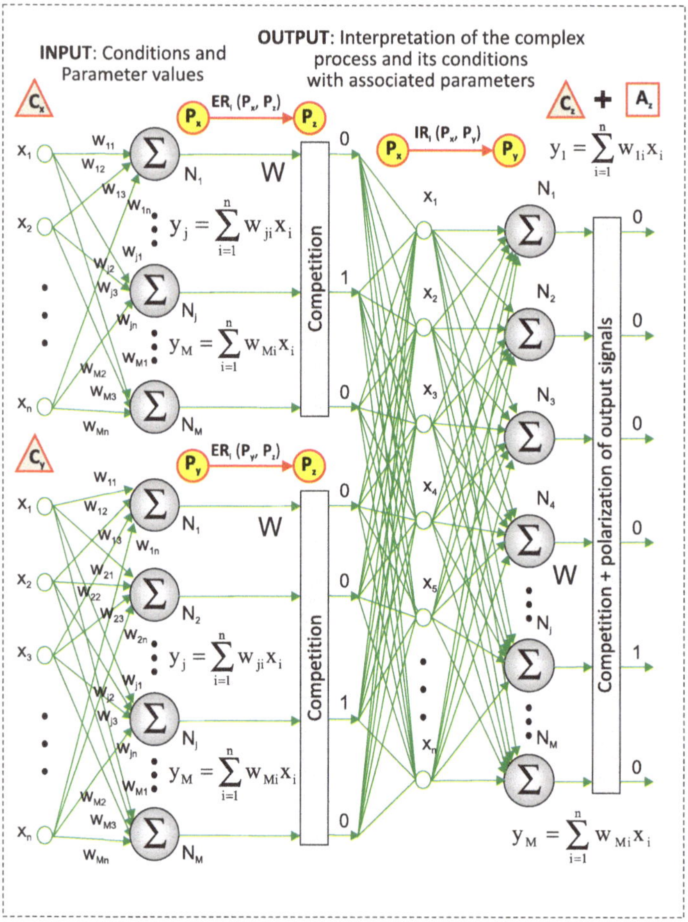

Fig. 10. A model of neural network enabling detection of antipatterns in operational configuration.

The control over the feature values of the machine elements decreases with time. This means that the access to basic measurements is reduced, and with it the ability to assess the fitness for function of its parts. We have observed several commercial attempts at automating detection of anomalies in operational data signals produced by machines. These include analysis of direct machine signals: vibrations (e.g. anomalies at specific frequencies for maintenance of ball bearings), power consumption (e.g. indicating incorrect engine operations), operational temperature, or other control metrics directly from the machine controllers (i.e. PLC). All of these approaches will benefit from explicit usage of antipatterns for example to define min/max boundaries, or to detect implicit data relations/structures producing machine failures.

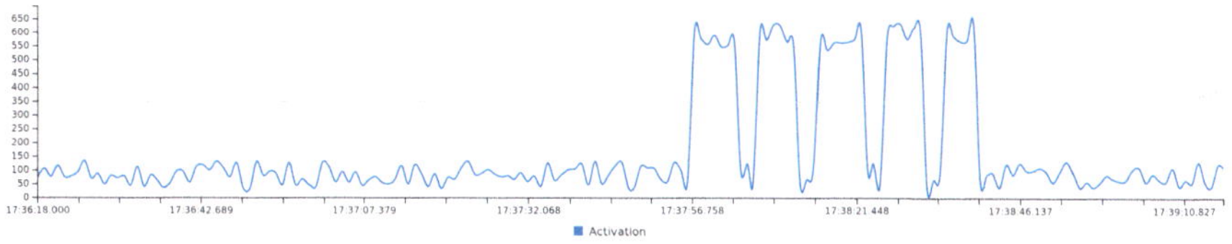

Fig. 11. A kinetic activation signal enabling antipattern identification in machine operations.

Antipatterns can also be automatically identified and used as a quality reference in analysis of indirect signals produced by the machine, like sound waves, overall magnitude of machine element's movement, or changes in the electromagnetic field. On Fig. 11, we present an example of dataset based on an indirect signal, with clearly visible five tasks performed by the machine element. Here automatic comparison of the data signal with antipattern library allows for automatic detection of defective elements, their wear, and replacement ordering. In this approach, detection of antipatterns could indicate a blocked tool (i.e. with extended low activation), excess forces (i.e. with singular peaks in activation signal).

6 Conclusions

Antipatterns are an optional, but important enhancement that can be added to quality assessment models in wide spectrum of applications in machine design, construction, and operations. As much as antipatterns have been always present in human expert knowledge, nearly infinite size of the antipattern data-set made it impossible to reflect on it using computers in the past. Recent advances in deep learning techniques, delivered new methods in automatically processing huge collections of multi-dimensional data, allowing antipatterns to become useful.

We have identified and validated antipatterns in all of phases of the machine's life-cycle (design, production, operation, recall) and separated them into classes: (1) value antipatterns (with simple error in feature values for a class of element), (2) relation based antipatterns (where error is in a composition of feature values), (3) structural antipatterns (with errors in the hierarchy of nodes decomposed by function). We have differentiated the classes in such a way due to their differences requiring changed approach to processing detecting them. The complexity of the antipattern class dictates the minimal complexity of models that have to be used to detect similarity to it. The antipattern feature values (1) can be identified using simple algebraic models, antipatterns found in feature relations (2) require an algorithmic model, and structural antipatterns (3) require deep learning models.

6.1 Further Research Opportunities

A shared library of antipatterns as set of complementary knowledge that would accompany element listings, would allow for a widespread adoption of the common knowledge on machine design, production, operations and recall. We foresee the main challenge here is to define a common, shared information exchange format with focus on describing fitness for a given function in a manner enabling automatic analysis.

Furthermore, detection of antipatterns in datasets created from measuring the indirect signals generated by the machine (e.g. vibrations, changes in electromagnetic field, temperature, electricity consumption) also provides cognitive value enabling failure detection, and could be further researched.

References

1. Koenig, A.: Patterns and antipatterns. J. Object-Oriented Program. **08**, 46–48 (1995)
2. Kacalak, W., Majewski, M., Budniak, Z.: Worm gear drives with adjustable backlash. J. Mech. Robot. **8**(1), 014504 (2015)
3. Kacalak, W., Majewski, M., Tuchołka, A.: Intelligent assessment of structure correctness using antipatterns. In: The Proceedings of the International Conference on Computational Science and Computational Intelligence CSCI 2015, Las Vegas, pp. 559–564. IEEE Xplore Digital Library (2015)
4. Kacalak, W., Majewski, M., Tuchołka, A.: A method of object-oriented symbolical description and evaluation of machine elements using antipatterns. J. Mach. Eng. **16**(4), 46–69 (2016)
5. Majewski, M., Kacalak, W.: Smart control of lifting devices using patterns and antipatterns. In: Silhavy, R., Senkerik, R., Kominkova Oplatkova, Z., Prokopova, Z., Silhavy, P. (eds.) CSOC 2017. Advances in Intelligent Systems and Computing, vol. 573, pp. 486–493. Springer, Cham (2017)
6. Kacalak, W., Majewski, M., Budniak, Z.: Innovative design of non-backlash worm gear drives. Arch. Civil Mech. Eng. **18**(3), 983–999 (2018)
7. Kacalak, W., Budniak, Z., Majewski, M.: Computer aided analysis of the mobile crane handling system using computational intelligence methods. In: Silhavy, R., Silhavy, P., Prokopova, Z. (eds.) CoMeSySo 2017. Advances in Intelligent Systems and Computing, vol. 662, pp. 250–261. Springer, Cham (2018)
8. Tuchołka, A., Majewski, M., Kacalak, W.: Object-oriented, symbolic notation for design features, relations and structures. Mach. Eng. **1**(20), 112–120 (2015)
9. Tuchołka, A., Majewski, M., Kacalak, W., Budniak, Z.: A method for intelligent quality assessment of a gearbox using antipatterns and convolutional neural networks. In: Silhavy, R. (ed.) CSOC 2018. Advances in Intelligent Systems and Computing, vol. 764, pp. 57–68. Springer, Cham (2018)
10. Tuchołka, A., Majewski, M., Kacalak, W., Budniak, Z.: Comparison of numerical models used for automated analysis of mechanical structures. In: Silhavy, R., Silhavy, P., Prokopova, Z. (eds.) CoMeSySo 2018. Advances in Intelligent Systems and Computing, vol. 859, pp. 341–352. Springer, Cham (2019)

Modelling of a Steel-Polymer Concrete Machine Tool Frame Component

Paweł Dunaj(✉), Stefan Berczyński, and Marcin Chodźko

West Pomeranian University of Technology Szczecin, Szczecin, Poland
{pawel.dunaj, stefan.berczynski, marcin.chodzko}@zut.edu.pl

Abstract. Designing machine tool bodies is a difficult task due to their complexity and requires a thorough knowledge of the phenomena accompanying the cutting process. It has therefore been a common practice to base design on proven generational structures, where the vast majority of machine tool bodies are assembled from materials such as steel or gray cast iron. More restrictive requirements impose the use of new materials for new constructions, related to increases in efficiency, accuracy and simultaneous reduction of energy consumption, as well as the level of noise emitted. There is a tendency to use composite materials characterized by low mass and very good damping properties. This article presents a method for modelling steel beams filled with a polymer concrete, as basic components of recently designed machine tool bodies. The modelling procedure was based on the finite element method. The accuracy of the proposed model was verified experimentally, resulting in less than 3% relative error in terms of natural frequencies. Next, on the basis of developed beam model, machine tool body consisting of steel beams filled with a polymer concrete material was modelled, resulting in 2.2% mean relative error. This paper presents the validity of the proposed model in allowing accurate prediction of the dynamic behaviour of machine tool bodies composed of steel beams filled with polymer concrete material. The presented method was used in the design process of a vertical lathe body.

Keywords: Finite element modelling · Machine tool · Composite beams · Modal analysis · Polymer concrete

1 Introduction

Dynamic properties are basic aspects considered in machine tool design process. Loads varying in time in the cutting process causes vibrations of various natures. In the vast majority of machine tools, this is an undesirable phenomenon, adversely increasing surface roughness, deterioration of dimensional and shape accuracy, and excessive tool wear. To avoid these phenomena, it is necessary to modify the process parameters, which often leads to a significant reduction in process efficiency. Therefore, anticipating machine tool dynamic and static properties at the design stage becomes pivotal [1–5].

The stability of the technological process, defined as the ability of the system (composed of the machine tool and the process itself) to return to equilibrium in the

M. Majewski and W. Kacalak (Eds.): IIRTS 2019, LNME, pp. 13–24, 2020.
https://doi.org/10.1007/978-3-030-37566-9_2

presence of small disturbances, is related to the dynamic characteristics of the load-bearing system of the machine tool, i.e. machine tool bodies, linear guides and drive systems [6, 7]. The biggest (in terms of dimension and inertia) and most complex elements in terms of geometry, are the machine tool bodies, an assembly of individual machine elements into a functional unit. Due to the complexity, the design process of these elements is a difficult task and requires in-depth knowledge of the phenomena accompanying the cutting process. It is therefore common practice to base design on previous proven solutions, introducing only minor changes when designing new machines.

Machine tool body elements are usually made from gray cast iron or a welded steel structure. Despite the well-established preferences for these materials, the restrictive requirements imposed on new constructions, desiring an increase in productivity and accuracy, and at the same time reductions in energy consumption, force new solutions, such as innovative hybrids combining materials such as steel, aluminium or gray cast iron with modern composite materials. Those are characterized by both good strength properties with relatively low weight and high vibration damping capabilities [8–10].

Do Suh et al. [11] proposed a hybrid solution consisting of a sandwich composite with outer layers of carbon fiber an aramid honeycomb core assembled with a steel profile. The developed components were then connected to a welded steel body by means of screwed and glued connections resulting in a significant improvement in damping properties and weight reduction. Kim and Chang [12] introduced a friction layer between aluminium-composite interfaces to increase the structural damping of the body parts of the machine tools, such as columns and spindle, tool holders, which resulted in reductions in mass and static deflection and much higher damping capabilities than those of aluminium constructions. Jung et al. [13] introduced composite aluminium hybrid beam structures with high-modulus carbon/epoxy composites to enhance dynamic stiffness and damping capacity of an inspecting machine for LCD glass panels.

Suh and Lee [11] presented the application of hybrid polymer concrete for precision machine tool beds. The hybrid polymer concrete bed was composed of welded steel structure and polymer concrete. It was designed for a high-speed gantry type milling machine through static and dynamic analyses using finite element method. The developed hybrid machine tool bed exhibited good damping characteristics over wide frequency range.

Sonawane and Subramanian [14] presented a milling machine body consisting of a grey cast iron, elastomer and concrete, aimed at increasing the damping of the structure. The evaluation of dynamic properties of the milling column was carried out with use of a finite element model. As a result, the damping properties have been improved by 20–30 times thanks to the introduction of a composite filling material, compared to unfilled ones.

The alternative solution presented in this paper is based on standard welded steel profiles with a square hollow cross-section, the interior of which is filled with a polymer concrete material composed of epoxy resin and filling fractions of various sizes [15, 16]. The advantage of this solution is the ability to shape the dynamics of the body by properly arranging the filling material, ensuring not only good damping properties but also the ability to increase the stiffness of the body elements at selected locations. This enables shaping both the structure and parameters of the mass-dissipative-elastic system over a wide range. Such a solution has a very high application potential not only in machine tool applications, but also in the case of assembly lines, vibratory machines and wherever machine work is accompanied by variable loads over time.

To appropriately distribute the filling material inside the body of a welded steel, tools to predict the dynamics of such a structure are required. Unfavourable coupling of spring-mass-damping properties can adversely affect the machine's efficiency and accuracy, therefore a properly designed machine tool load-bearing system is needed to determine its functionality [17, 18]. Due to the lack of existing modelling procedures for machine tool bodies made using such a technology, it was necessary to develop a specific method. To achieve this objective, modelling procedures and experimental verifications on 50 × 50 mm beam was conducted. Next, based on these established beam model, a one third scale model of machine body was built.

2 Concept of Hybrid Machine Tool Body

The presented solution is based on a welded steel structure filled with material with different filling fraction sizes. Steel welded construction ensures adequate stiffness, while polymer concrete provides high damping abilities. Hybrid body composed of steel beams filled with polymer concrete material is schematically shown in Fig. 1.

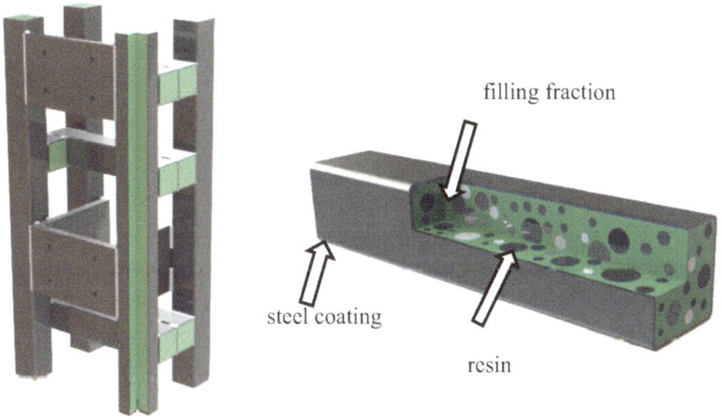

Fig. 1. Machine tool body composed of steel beams profiles filled with polymer concrete material.

The research object was a steel beam with a length of 1000 mm, square cross-section of 50 × 50 and wall thickness 2 mm. The analyzed beam was filled with a polymer concrete.

The polymer concrete consisted of different sizes of filling fraction in an epoxy resin binder. The mineral filling consisted of four fractions: ash, a small fraction of mostly sand with a grain size 0.25–2 mm, medium fraction (2–10 mm) and coarse fraction (8–16 mm) of mostly gravel. The medium and coarse fraction grain shapes were irregular. The mass percentage of the subsequent mineral fillings is presented in Table 1.

Table 1. Composition of the polymer concrete (mass parentage).

	Resin	Ash	Small fraction (0.25–2 mm)	Medium fraction (2–10 mm)	Coarse fraction (8–16 mm)
Polymer concrete material	15%	1%	19%	15%	50%

The polymer concrete was prepared as follows. First, the individual fractions were mixed in the appropriate proportions. Epoxy resin was then added to the prepared compound. Subsequently, the hardener in proportions recommended by the manufacturer was added. To evenly distribute all ingredients, the prepared compound was mixed thoroughly. After that, the steel profiles were filled. The polymer concrete could cure for 48 h.

3 Finite Element Model

To anticipate the dynamic properties of the machine tool body composed of steel profiles filled with polymer concrete, a model of a single 50 × 50 mm beam was built using the finite element method. A geometrical model was discretized using a Midas NFX preprocessor [19]. Both the steel outer and polymer concrete filling were discretized using six-sided isoparametric solid elements with 8 nodes (CHEXA) and five-sided isoparametric solid element with 6 nodes (CPENTA), also known as wedge elements.

To improve the efficiency of the FEM, a structured meshing technique was applied. Next, to avoid accuracy degradation, mesh quality was analyzed both in terms of aspect ratio and skewness. The contact between the steel outer and the polymer concrete was modelled using coincidental nodes. The use of such a model was chosen from research conducted on concrete filled steel tubes (CFST). It can be concluded that both the steel outer and the polymer concrete filling transfer the load behaving like beams made of one material [20–24]. In total, the model of a single beam was composed of 19,600 elements and 21,985 degrees of freedom. The discretized model is shown in Fig. 2.

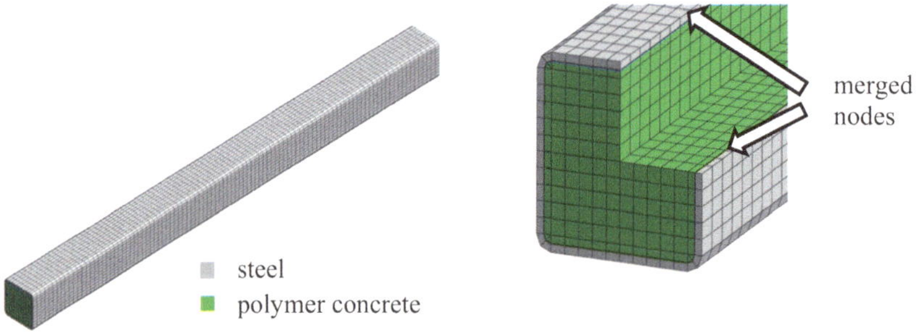

Fig. 2. Discretized beam model.

Regardless, the character of the polymer concrete (the occurrence of filling fractions of various size combined with epoxy resin), owing to preliminary tests (i.a. verification of Maxwell's Reciprocity Theorem, static compression tests, experimental mode shapes analysis) it was decided, that in the range of the analyzed displacements (up to 0.5 mm – the maximum displacement of machine tool vibrating body elements does not usually exceed this value [25, 26]), the polymer concrete can be modelled as a linear isotropic material, primarily considering the material properties of the epoxy resin.

To determine the Young's modulus and Poisson ratio, required for FEM modelling, static tests were performed. These were carried out on an Instron 8850 machine working in an air-conditioned laboratory with at a temperature of 23 °C and relative humidity of 50%. Before testing, the samples spent 72 h acclimatising to the air-conditioned laboratory. Table 2 contains material data determined from the static tests and the damping ratio, which was determined from frequency response functions (FRFs) using a half power method.

Table 2. Material properties.

Parameter	Steel	Polymer concrete
Young's modulus	210 ± 5 GPa	16.8 ± 0.2 GPa
Poisson's ratio	0.28 ± 0.03	0.20 ± 0.05
Density	7487 ± 35 kg/m^3	2118 ± 6 kg/m^3
Damping ratio ζ	0.0011 ± 0.00005	0.0024 ± 0.00012

To describe the damping of the structure a complex stiffness [27, 28] model was used.

4 Experimental Model Validation

To verify the established model an impact test was performed. The layout of the test stand is shown in Fig. 3. To approximate free boundary conditions, the tested beam was suspended on steel cables. The experiment was performed using Siemens TestLab software and Scadas III hardware and included i.a. data processing, monitoring of power spectral density and coherence functions. Excitation was carried out using a PCB 086C01 modal hammer in two perpendicular axes Z and X. Measurement of the response was carried out at 56 points on the X and Z axes using three-axis PCB 356A01 ICP accelerometers with a sensitivity of 10 mV/m/s^2. Double impacts or overload of any channel result with automatic rejection of measurement.

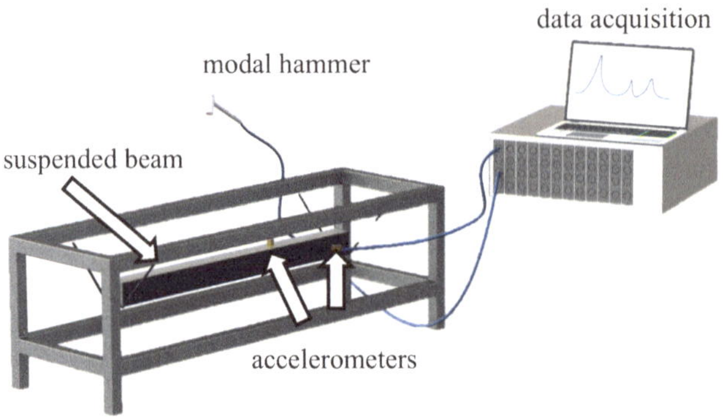

Fig. 3. Modal analysis test stand.

As a result of the experiment, 56 frequency response functions were determined using H1 type of FRF estimator. The poles of the modal model were estimated using the Polymax algorithm. The estimation process was supported by a stabilization diagram interpretation and a preliminary version of the modal model was built. Before the final interpretation the model was validated using an MAC indicator. Natural frequencies and mode shape sets for the examined beam were presented in Table 3 and Fig. 4 respectively. Relative error δ was determined as follows:

$$\delta = \left| \frac{\omega_{exp} - \omega_{fem}}{\omega_{exp}} \right| \cdot 100\% \tag{1}$$

where: ω_{exp} – natural frequency obtained from the experiment; ω_{fem} – natural frequency obtained from the FEM analysis.

Table 3. Comparison of natural frequencies between FEM model and experimental results.

Mode number	FEM results	Experimental results	Relative error δ
1	248 Hz	247 Hz	<1%
2	248 Hz	251 Hz	~1%
3	685 Hz	676 Hz	~1%
4	685 Hz	678 Hz	~1%
5	1241 Hz	1276 Hz	<3%

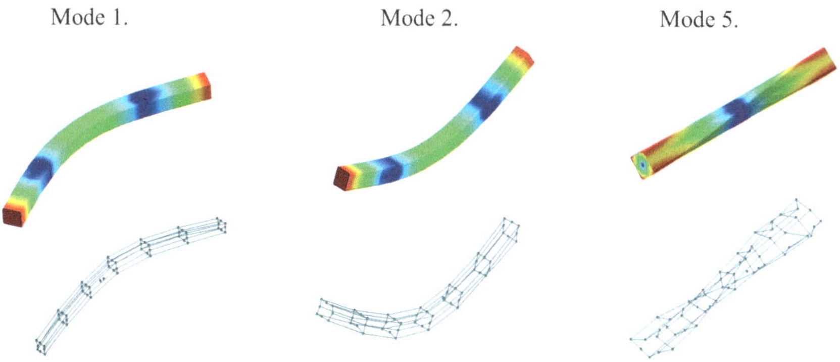

Fig. 4. Selected mode shapes comparison for beam with 50 × 50 cross-section.

Figure 5 contain frequency response function comparison for the beam.

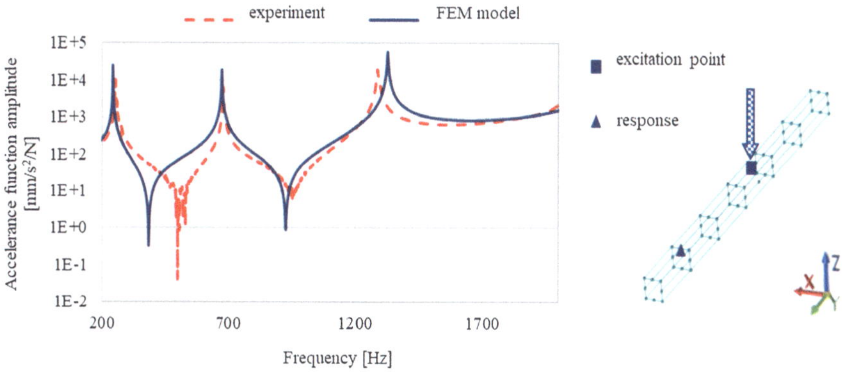

Fig. 5. Frequency response function comparison for 50 × 50 beam.

The developed procedure and identified parameters allow for development of reliable model for steel-polymer concrete beam with cross-Section 50 × 50 mm (relative error δ not exceeding 3%, on average 1.4%).

5 Beam Model Application in Machine Tool Body Modelling

This chapter presents an example of the developed model for determining the dynamics of a machine tool body consisting of hollow steel profiles with a square cross-section of 50 × 50 mm and wall thickness 2 mm filled with polymer concrete. The analyzed body is a slightly modified reflection of the one third scale of the newly designed light construction vertical lathe.

Construction of the finite element model began by defining the geometrical model, with necessary simplifications comprising removing non-essential features from dynamical properties of the machine tool viewpoint, i.e. small holes, small roundings, chamfers, etc. After preparing the geometric model, discretization was made using CHEXA and CPENTA elements (Fig. 6). In total, the model of the machine tool body was composed of 60,584 elements and 225,384 degrees of freedom. Support was modelled using a beam elements (CBAR) with an equivalent stiffness and damping.

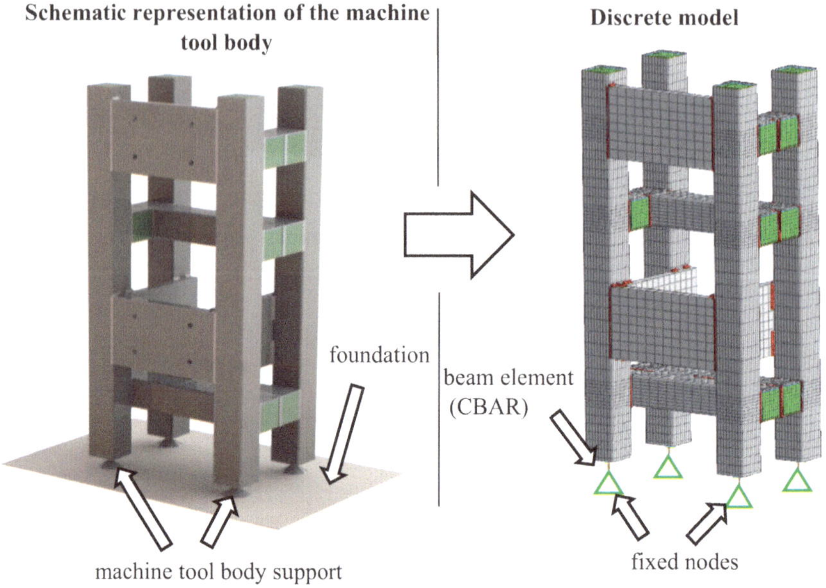

Fig. 6. Geometric model subjected to the process of discretization.

Modal analysis of the discretized model was conducted using NEi Nastran Solver (SOL103). Table 4 and Fig. 7 contain comparisons of the natural frequencies and mode shapes obtained from the FEM analysis and experimental testing.

Table 4. Comparison of natural frequencies between FEM model and experimental results of a machine tool body composed of 50 × 50 mm steel beams filled with polymer concrete.

Mode number	FEM results	Experimental results	Relative error δ
1	437 Hz	431 Hz	<1%
2	858 Hz	791 Hz	<9%
3	936 Hz	928 Hz	<1%
4	957 Hz	969 Hz	~1%
5	1032 Hz	1041 Hz	<1%
6	1141 Hz	1151 Hz	<1%
7	1195 Hz	1232 Hz	~3%

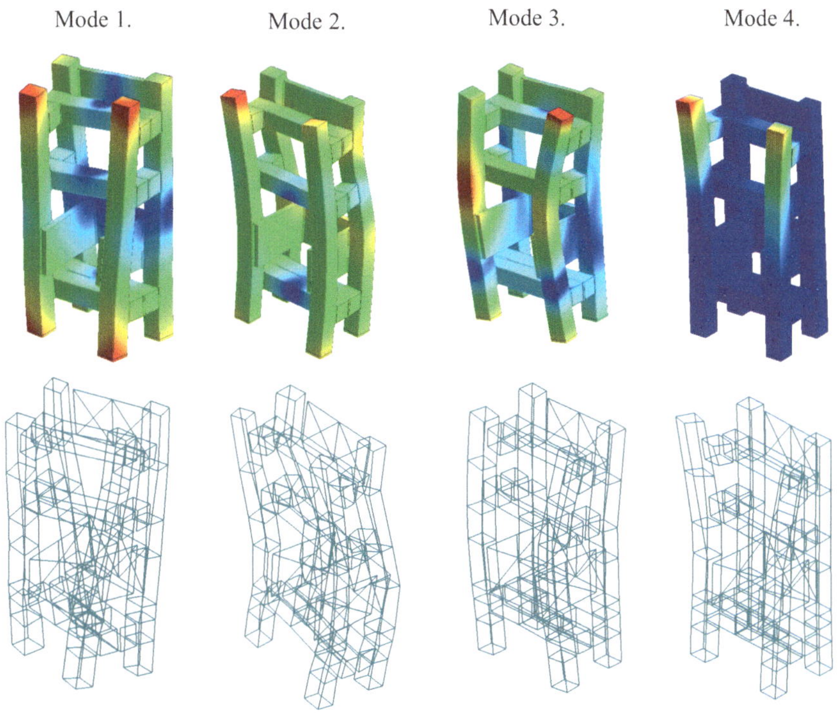

Fig. 7. Comparison of selected mode shapes between FEM model and experimental results for machine tool body.

Figures 8 and 9 present FRF comparisons at selected points, where the tool is mounted and where the spindle is mounted.

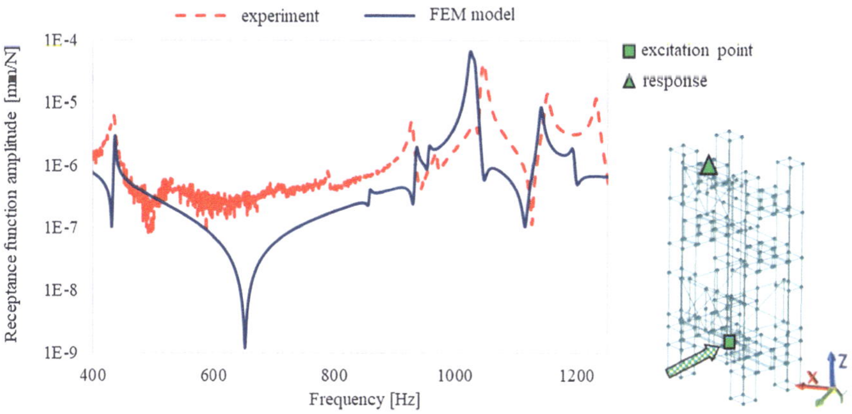

Fig. 8. Receptance function comparison for FEM model and experimental data, between excitation point and point where tool is mounted in Y direction.

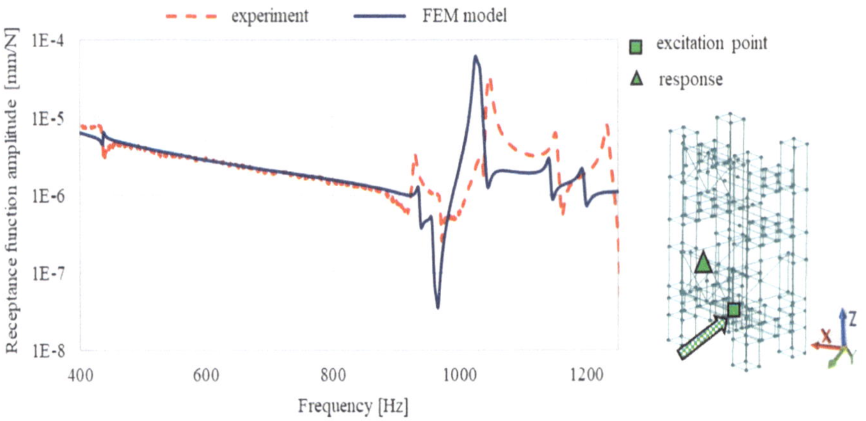

Fig. 9. Receptance function comparison for FEM model and experimental data between excitation point and point where spindle is mounted in Y direction.

Summing up, it can be stated that in the case of the modelled machine tool body, the applied modelling methodology allows obtaining results with a relatively high accuracy. The value of the relative error δ for the natural frequencies did not exceed 9%, on average 2.2%. The differences in the second natural frequency may result from inaccuracies in the determination of support equivalent stiffness.

Comparing selected receptance functions, their high compatibility was noticed (excluding the mentioned frequency), this indicates the good ability of the model to predict the dynamic properties of the machine tool body in the analyzed frequency range.

6 Conclusions

Accurate prediction of the dynamic properties of a machine tool bodies in the design phase significantly reduces the cost of developing the prototypes of a new solution. Having an appropriate modelling procedure also allows shaping of these properties and selection of the appropriate components for the designed system.

Such a high accuracy of modelling was obtained thanks to a series of identification tests of components exhibiting various proportions (ratios) of geometric dimensions, and application of the correct methodology for modelling welded connections of these elements. The dimensions of the components were selected considering the target dimensions of the prototype of a lightweight construction lathe.

Differences of just 2–3% between the natural frequencies corresponding to individual mode shapes and the compatibility of frequency response functions, confirm the correctness of the developed modelling methodology for the structural component of machine tool body.

On the basis of the developed component model, the machine tool body model was built achieving high compatibility with experiment.

Acknowledgements. The research work reported in this paper was made possible by the EU grant: "Light construction vertical lathe" POIR.04.01.02-00-0078/16.

References

1. Altintas, Y.: Manufacturing Automation: Metal Cutting Mechanics, Machine Tool Vibrations, and CNC Design. Cambridge University Press, Cambridge (2012)
2. Koenigsberger, F., Tlusty, J.: Machine tool structures. Elsevier (2016)
3. Pajor, M., Marchelek, K., Powalka, B.: Experimental verification of method of machine tool-cutting process system model reduction in face milling. WIT Trans. Model. Simul. **22** (1999)
4. Pajor, M., Marchelek, K., Powałka, B.: Method of reducing the number of DOF in the machine tool-cutting process system from the point of view of vibrostability analysis. Modal Anal. **8**(4), 481–492 (2002)
5. Jasiewicz, M., Miadlicki, K., Powałka, B.: Assistance of machining parameters selection for slender tools in CNC control. In: AIP Conference Proceedings, vol. 2029, p. 020024 (2018)
6. Chlebus, E., Dybala, B.: Modelling and calculation of properties of sliding guideways. Int. J. Mach. Tools Manuf **39**(12), 1823–1839 (1999)
7. Pawełko, P., Berczyński, S., Grządziel, Z.: Modeling roller guides with preload. Arch. Civil Mech. Eng. **14**(4), 691–699 (2014)
8. Angus, H.T.: Cast iron: physical and engineering properties. Elsevier (2013)
9. Bhandari, V.B.: Design of Machine Elements. Tata McGraw-Hill Education, New York (2010)
10. Smithells, C.J.: Metals reference book. Elsevier (2013)
11. Do Suh, J., Kim, H.S., Kim, J.M.: Design and manufacture of composite high speed machine tool structures. Compos. Sci. Technol. **64**(10–11), 1523–1530 (2004)
12. Kim, J.-H., Chang, S.-H.: Design of μ-CNC machining centre with carbon/epoxy composite–aluminum hybrid structures containing friction layers for high damping capacity. Compos. Struct. **92**(9), 2128–2136 (2010)
13. Jung, S.C., Lee, J.E., Chang, S.H.: Design of inspecting machine for next generation LCD glass panel with high modulus carbon/epoxy composites. Compos. Struct. **66**(1–4), 439–447 (2004)
14. Sonawane, H., Subramanian, T.: Improved dynamic characteristics for machine tools structure using filler materials. Procedia CIRP **58**, 399–404 (2017)
15. Dunaj, P., Okulik, T., Powałka, B., Berczyński, S., Chodźko, M.: Experimental investigations of steel welded machine tool bodies filled with composite material. In: Gapiński, B., Szostak, M., Ivanov, V. (eds.) MANUFACTURING 2019. LNME, pp. 61–69. Springer, Cham (2019)
16. Okulik, T., Dunaj, P., Chodźko, M., Marchelek, K., Powałka, B.: Determination of dynamic properties of a steel hollow section filled with composite mineral casting. In: Gapiński, B., Szostak, M., Ivanov, V. (eds.) MANUFACTURING 2019. LNME, pp. 561–571. Springer, Cham (2019)
17. Berczyński, S., Gutowski, P.: Identification of the dynamic models of machine tool supporting systems. part i: an algorithm of the method. J. Vibr. Control **12**(3), 257–277 (2006)
18. Gutowski, P., Berczyński, S.: Identification of the dynamic models of machine tool supporting systems. Part II: an example of application of the method. J. Vibr. Control **12**(3), 279–295 (2006)
19. Midas, I.T.: User's Manual of midas NFX. MIDAS IT (2011)

20. Dai, X.H., Lam, D., Jamaluddin, N., Ye, J.: Numerical analysis of slender elliptical concrete filled columns under axial compression. Thin-Walled Struct. **77**, 26–35 (2014)
21. Duarte, A.P.C., Silva, B.A., Silvestre, N., De Brito, J., Júlio, E., Castro, J.M.: Finite element modelling of short steel tubes filled with rubberized concrete. Compos. Struct. **150**, 28–40 (2016)
22. Hu, H.-T., Huang, C.-S., Chen, Z.-L.: Finite element analysis of CFT columns subjected to an axial compressive force and bending moment in combination. J. Constr. Steel Res. **61** (12), 1692–1712 (2005)
23. Li, S., Han, L.-H., Hou, C.: Concrete-encased CFST columns under combined compression and torsion: analytical behavior. J. Constr. Steel Res. **144**, 236–252 (2018)
24. Ren, Q.-X., Han, L.-H., Hou, C., Tao, Z., Li, S.: Concrete-encased CFST columns under combined compression and torsion: experimental investigation. J. Constr. Steel Res. **138**, 729–741 (2017)
25. Chodźko, M.: The search for weak elements affecting the vibrostability of the system consisting of the machine tool and the cutting process, based on symptoms observed during operation. Exploit. Probl. **1**, 5–23 (2015)
26. Chodźko, M., Marchelek, K.: Experimental tests of dynamic properties of machine tool body systems. Sel. Issues Mach. Eng. **16**, 67–81 (2011). (in Polish)
27. Holanda, S.A., Silva, A.A., de Araújo, C.J., de Aquino, A.S.: Study of the complex stiffness of a vibratory mechanical system with shape memory alloy coil spring actuator. Shock Vibr. **2014** (2014). Articleid. 162781
28. Neumark, S.: Concept of Complex Stiffness Applied to Problems of Oscillations with Viscous and Hysteretic Damping. HM Stationery Office, London (1962)

Machine Learning Based Heuristic Technique for Multi-response Machining Process

Tamal Ghosh(✉) and Kristian Martinsen

Department of Manufacturing and Civil Engineering,
Norwegian University of Science and Technology,
Teknologivegen 22, 2815 Gjøvik, Norway
{tamal.ghosh, kristian.martinsen}@ntnu.no

Abstract. Manufacturing process variables influence the quality of products substantially. It is unquestionably difficult to model the manufacturing processes that include a large number of variables and responses. Development of the multi-objective surrogate models for the manufacturing processes could be computationally and economically expensive. In this article, a generic multi-objective surrogate-coupled heuristic algorithm is employed that needs small amount of experimental data as input, and predicts precise responses with quick Pareto solutions. The proposed algorithm is verified with different cases collected from the literature based on the CNC turning, centerless cylindrical grinding, and micro milling machining and shown to produce some interesting results.

Keywords: Manufacturing process optimization · Surrogate models · Heuristic algorithm · Multi-objective optimization

1 Introduction

Manufacturing process optimization has gained substantial interest from the researchers since decades. In reality, it is difficult to optimize the process of manufacturing due to the association of a large number of design parameters and multiple objectives or performance indicators. Multidisciplinary collaborative approaches are required to manufacture complex engineering components. It also requires complicated design space due to the nonlinear relationships among various dependent and independent design variables to manufacture these optimally. Due to the above fact, it is substantially difficult to formulate these relationships mathematically [1]. Precision of manufacturing process largely depends on the empirical process data, which is expensive in terms of the machining costs, labor costs, overhead costs, and scrap costs. Data-driven models, machine-learning techniques, meta-model or surrogate modeling techniques could be appropriate in such scenarios [2].

The complexity of engineering product increases with the number of design variables, performance indicators and specifications or tolerance levels defined for the products. This problem is termed as combinatorial problem, which could have many near optimal solutions within boundary conditions. For an example, the milling machining is associated with various design parameters such as the spindle speed, feed

M. Majewski and W. Kacalak (Eds.): IIRTS 2019, LNME, pp. 25–35, 2020.
https://doi.org/10.1007/978-3-030-37566-9_3

rate, depth of cut, tool diameter, etc. while minimizing the surface roughness, applied cutting forces, tool wear, which in turn improve the material removal rate and tool life. However, this optimization could be very complex due to the nonlinear relationships among the input and output variables. Moreover, this kind of optimization approach is substantially process specific and generic formulations or models are not prominently available in the literature [3]. Hence, the aim of this study is to portray a suitable multi-objective process modeling to solve the above problem promptly.

To cater the purpose, an Artificial Neural Network (ANN) and Gaussian Kernel Regression (GKR) techniques are considered to construct efficient surrogate models, which take various design parameters as the inputs. Once the surrogates are trained, these are capable of predicting process responses with good accuracy. Both the surrogates studied here, are compared with each other based on the Normalized Mean Square Error (NMSE) and the best one is selected.

In general, Design of Experiment (DOE) is used to define the design space for the machining or manufacturing processes [4]. For that purpose, the Latin hypercube sampling (LHS) [5] is used, which uniformly generates the set of values for design parameters within the predefined range. Thereafter a heuristic method is employed to select optimal sets of these parameters, which could fine-tune the process responses.

2 Manufacturing Process Optimization

CNC (Computerized Numerical Control) machining has transformed the manufacturing industries drastically since past few decades, which precisely removes excess materials with precision and accuracy while improving the tool life and the quality of the manufactured products with reduced manufacturing cost. This includes the turning, grinding, milling, welding, lathe, which effectively obtain flat or curved shaped components with smooth finishing [6]. In this paper, three cases are considered based on the CNC turning, grinding, and milling operations. In CNC turning, design parameters such as the cutting tool geometry and materials, the depth of cut, feed rates, cutting speeds as well as the use of cutting fluids could influence the material removal rates and the other performance indicators, such as the surface roughness, roundness of circular and dimensional deviations of the product [7]. Whereas, the cylindrical grinding is employed to finish parts with circular symmetry and projections, variations in shapes, varying diameters etc. In in-feed method, the shape variations are accommodated in the form of the grinding wheel aiming to form various part diameters and lengths to meet the part geometry [8]. In this process, the part is fed to the wheels from the above with no lateral movement of the piece while it is being on ground. However, in milling process, machining at the top speeds usually attains better surface roughness, material removal rate, and improved tool wear rate. The effect of the input parameters such as the cutting speed, feed per tooth and depth of cut for milling could be decisive factors for the product quality in a big way [9].

For manufacturing process optimization, the experimental design approaches are heavily practiced in literature, which are active statistical method. The design space is formed with a DOE tool, which makes small changes in one or more inputs to analyze the corresponding influences in the response patterns. This analysis could essentially points out the complex correlations among the process parameters and responses, which further reduces the process variabilities [6]. This experimental design space could be useful for evaluating the nature of the manufacturing process and considered as the input to the process optimization models. Various methods are available in the literature, which are categorized as the DOE techniques. These are full factorial method, fractional factorial method, Taguchi's method, LHS etc. Various applications of the DOE based approaches to the manufacturing process optimizations could be found in the literature for the single or multiple performance indicators [10]. In this study, three different cases are collected from the literature [7–9], which are used as the training data for the proposed iterative heuristic optimizer. The cases are discussed in the following subsections.

2.1 CNC Turning Case Study

The experimental data for CNC turning are collected from the literature and presented in Table 1. A rigid CNC turning center with a 7.5 kW spindle motor at 4200 rpm (machine type of Vtum–20, manufactured by VICTOR Taichung Machinery Works Co. in Taiwan) is used to perform machining [7].

Table 1. CNC turning process data [7].

No.	CS	FR	DC	CFMR	Ra	Rt	phi
1	125	0.12	0.5	4	3.27	28.76	4.7
2	125	0.16	0.65	8	1.92	15.91	2.9
3	125	0.2	0.8	12	1.8	11.04	1.9
4	155	0.12	0.65	12	1.11	6.626	0.85
5	155	0.16	0.8	4	1.78	7.861	1.1
6	155	0.2	0.5	8	2.73	10.84	1.95
7	185	0.12	0.8	8	1.04	5.698	0.75
8	185	0.16	0.5	12	2.87	11.19	2.1
9	185	0.20	0.65	4	3.91	15.71	2.55

The cutting length of the work piece is 40 mm. The cutting tool is made of carbide and coated with titanium nitride (TiN) manufactured by Sumitomo Electric Industries, Ltd., Japan with a part number of TNMG160408-UG. In this case, the cutting speed (CS) (100–200 m/min), feed rate (FR) (0.1–0.25 mm/rev), depth of cut (DC) (0.1–1.0 mm) and the cutting fluids mixture ratio (CFMR) (2–15%) are considered as most influential parameters of turning operations and average surface roughness (Ra), maximum surface roughness (Rt) and roundness of circumference surface (RCS) are considered as process responses. The material considered is SKD11 (JIS) alloy tool steel with high carbon high chromium. This is used in the production of dies, plastic injection molding dies, precision gauge, spindle, jigs and fixtures, etc. The composition of the SKD11 is 1.55 wt.% C–11.5 wt.% Cr–0.70 wt.% Mo–1.00 wt.% V–0.30 wt.% Mn–0.25 wt.% Si.

The yielding stress of raw SKD11 is 330 MPa, the Young's modulus is 200 GPa and hardness is 25 HRC. The diameter for the work pieces has been fixed to 20 mm.

2.2 Centerless Cylindrical Grinding Case Study

An in-feed centerless cylindrical grinding machine is considered for the experiments [8]. The material considered for the experiments is made of EN52 for an internal combustion (IC) engine valve stem with 79.6 mm dia. An A80N5V45 grinding wheel rotating at 1440 rpm (giving a surface speed of 45 m/s) and an A80RR control wheel were utilized for this process. The centerless cylindrical grinding process was carried out over a length of 98 mm of valve stem with a job height of 212 mm above the blade. The chemical composition of the EN52 is given as, C (0.40–0.50%) Si (2.70–3.30%) Mn (0.8% max) P (0.04% max) S (0.03% max) Cr (8–10%). For centerless cylindrical grinding process, surface roughness (Ra), out of cylindricity (OC) and diametral tolerance (Dt) were selected as response variables whereas dressing feed (DF) (2–15 mm/min), grinding feed (GF) (1–12 mm/min), dwell time (DT) (1–5 s) and cycle time (CT) (5–15 s) are considered as experimental design parameters. Table 2 presents the experimental data for the cylindrical grinding process.

Table 2. Cylindrical grinding process data [8].

No.	DF	GF	DT	CT	Ra	OC	Dt
1	5	2	1.5	10	0.432	0.67	0.001
2	5	6	2.5	11	0.439	0.67	−0.001
3	5	10	3.0	12	0.427	1	−0.002
4	8	2	2.5	12	0.578	1.67	0.001
5	8	6	3.0	10	0.613	1	0.001
6	8	10	1.5	11	0.763	1.33	0.001
7	10	2	3.0	11	0.505	3.33	0.004
8	10	6	1.5	12	0.517	3	0.004
9	10	10	2.5	10	0.554	3.67	0.003

2.3 CNC Micro-milling Case Study

The micro-milling experimental data are collected from the published paper and portrayed in Table 3. The experiments were carried out using a DECKEL MAHO DMU 60 PCNC milling machine [9].

Al7075 material is used (Vickers hardness of 139) as a work piece material, which had a dimension of $15 \times 10 \times 20$ mm^3. The chemical compositions of material are given as, Li < 0.0002 wt%, Si 0.92 wt%, Mn 0.348 wt%, P < 0.001 wt%, Sr < 0.0001 wt%, Cr 0.093 wt%, Ni 0.057 wt%, Na 0.003 wt%, Al 89.0 wt%, Cu 1.71 wt%, Co < 0.001 wt%, Ti 0.048 wt%, Be 0.0003 wt%, V 0.009 wt%, Fe 0.55, Pb wt%, 0.018 wt%, Mg 2.00 wt%, B 0.0017 wt%, Sn 0.008 wt%, Zn 5.22wt%, Ag 0.0022 wt%, Bi 0.0018 wt%, Ca 0.0027 wt%, Cd 0.0031 wt%, and Zr 0.0078 wt%. In this case, Spindle speed (SS) (5000–15000 rpm), feed per tooth (FPT) (0.5–1.5 μm/tooth) and

Table 3. Taguchi's design sets for micro milling process [9].

No.	SS	FPT	DC	TW	Fx	Fy	Ra
1	10000	0.5	50	5.41	1.33	0.71	0.33
2	10000	1.0	75	13.51	1.63	0.86	0.47
3	10000	1.5	100	16.22	2.02	1.02	0.71
4	11000	0.5	75	27.02	1.62	1.08	0.32
5	11000	1.0	100	32.43	2.03	1.49	0.59
6	11000	1.5	50	14.86	2.65	1.52	0.58
7	12000	0.5	100	45.95	2.10	1.36	0.27
8	12000	1.0	50	27.03	2.99	1.66	0.35
9	12000	1.5	75	32.43	3.55	1.81	0.58

depth of cut (DC) (50–100 μm) were considered as design parameters and tool wear (TW), cutting forces (Fx, Fy), and surface roughness (Ra) were selected as process responses.

3 Iterative Surrogate-Assisted Heuristic

The Taguchi's design based data sets are employed for the training of the surrogate functions for the proposed heuristics. All these data contain the input data and output responses. Following surrogate functions are used in this work.

3.1 Artificial Neural Network (ANN)

In order to construct the ANN based surrogate model, a Cascade-Forward Neural Network (CFNN) is considered, which consists of the input layer, hidden layer, and output layer [11]. The number of neurons in the input layer and the output layer depends on the number of design parameters and the performance indicators. The CFNN has some direct connections among the inputs and outputs on top of the multi-layer feed forward perceptron architecture. The response equation for the CFNN is given in Eq. (1) as,

$$y_i = \sum_{k=1}^{n} Z_i^k \times w_j^k x_k + Z_i^{oa}\left(\sum_{j=1}^{m} w_{ji}^{oa} \times Z_k^{ha}\left(\sum_{k=1}^{n} w_{jk}^{ha} x_k\right)\right) \tag{1}$$

$$y_i = \sum_{k=1}^{n} Z_i^k \times w_j^k x_k + Z_i^{oa}\left(\beta_i + \sum_{j=1}^{m} w_{ji}^{oa} \times Z_k^{ha}\left(\beta_j + \sum_{k=1}^{n} w_{jk}^{ha} x_k\right)\right) \tag{2}$$

Where Z_i^{oa} is denoted as the activation function for i^{th} output y_i, w_{ji}^{oa} is the weight from j^{th} hidden layer neuron to i^{th} output node, Z_k^{ha} is the activation function for j^{th} hidden layer neuron, w_{jk}^{ha} is the weight from k^{th} input to j^{th} hidden layer neuron, and x_k is the kth input signal. Further, if some bias is added to input layer, the Eq. (1) becomes

Eq. (2), Z_i^k is the activation function and w_j^k is the weight from the inputs to outputs. The network weight in the CFNN is approximated based on the neurons in the input layer.

3.2 Gaussian Kernel Regression (GKR)

The GKR is based on the data mapping from the low-dimensional space to high-dimensional space. This linear regression model in high-dimensional space is equivalent to the Gaussian regression in the low-dimensional space. The linear regression learner is based on the Support Vector Machine (SVM) regression. In this approach, the input parameters x are mapped onto an m-dimensional attribute space using the non-linear mapping, which further converts it to a linear model in the same attribute space. The linear model is,

$$f(x, \omega) = \sum_{j=1}^{m} \omega_j g_j(x) + \beta \tag{3}$$

Where $g_j(x)$ $(j = 1 \ldots m)$ is a non-linear transformation function and β is the bias. The performance of regression is analyzed using the ε-insensitive loss function [12],

$$L(y, f(x, \omega)) = \begin{cases} 0 & if |y - f(x, \omega)| \leq \varepsilon \\ |y - f(x, \omega)| - \varepsilon & Otherwise \end{cases} \tag{4}$$

The above mentioned regression model can be transformed into an optimization problem using,

$$\min Z = \frac{1}{2}\|\omega^2\| + C\sum_{k=1}^{n}(\gamma_i + \gamma_i^*) \tag{5}$$

Subject to,

$$\begin{cases} y_i - f(x_i, \omega) \leq \varepsilon + \gamma_i^* \\ f(x_i, \omega) - y_i \leq \varepsilon + \gamma_i \\ \gamma_i, \gamma_i^* \geq 0, i = 1 \ldots n \end{cases} \tag{6}$$

Where γ_i and γ_i^* $(i = 1 \ldots n)$ are positive slack variables, which can calculate the deviation of the input parameters beyond the ε-insensitive neighborhood. This optimization problem is known as the *primal*. It could be transformed into a *dual* and an exact SVM kernel is required to solve this. Whereas, in SVM assisted GKR, the *primal* is solved using the high-dimensional attribute space [13].

3.3 Normalized Mean Square Error (NMSE)

In this study, the NMSE is employed for the performance evaluations of the surrogates. The NMSE is a metric that measures the overall deviations between the predicted and measured values. It is defined as,

$$NMSE = \frac{1}{N}\sum_i \frac{(P_i - M_i)^2}{\vec{P}\,\vec{M}} \tag{7}$$

$$\vec{P} = \frac{1}{N}\sum_i P_i \tag{8}$$

$$\vec{M} = \frac{1}{N}\sum_i M_i \tag{9}$$

Where P is the predicted output, M is the measured output and N is the number of observations. In NMSE computation, the deviations (absolute values) are summed instead of the differences. Due to that fact, the NMSE generally shows the most striking differences among the models. If a model has a very low NMSE score, then it is an optimal model in the space and time.

Once both the surrogates are trained, the comparison is done based on the obtained NMSE scores and R-values obtained for the regression error. From Table 4, it can be stated that the GKR based approach outperforms the ANN model for all the training data. Therefore, the GKR surrogate model is selected, which is further used as a fitness function to the proposed heuristic. Figure 1 depicts the ANN based regression plots for the case data with promising R-values.

Table 4. ANN vs GKR: comparison based on NMSE values

Cases	GKR		ANN	
	NMSE	R-value	NMSE	R-value
CNC turning operation	**0.0487**	**0.93704**	0.8444	0.84361
Cylindrical grinding	**0.0474**	**0.96721**	0.3079	0.87762
CNC micro-milling	**0.0124**	**0.98418**	0.8701	0.95655

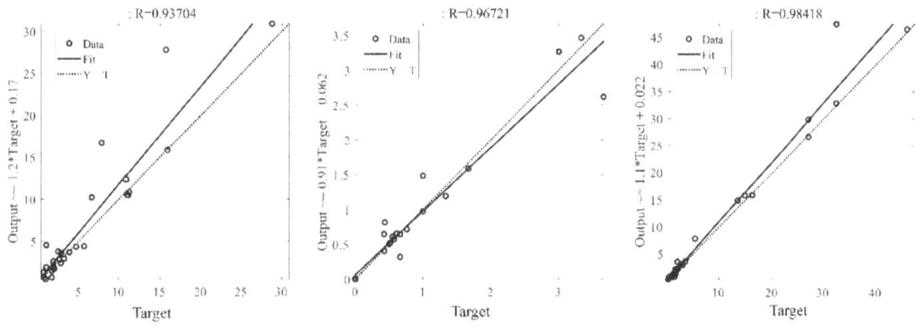

Fig. 1. Regression plots for three cases.

3.4 Heuristic Algorithm

The proposed search technique is an improvement heuristic, which is population based and iterative in nature. The steps of the proposed algorithm are as follows,

Step 1. It generates a new population at the initial iteration using the LHS based algorithm, which uniformly generates the design space for the input parameters. Every row of the population table is an experimental setup. The size of the initial population is set to 100. Maximum number of iterations are set to 100.
Step 2. The GKR based fitness function is used for functional evaluation, which predicts the output for each of the experiment setup (population member) obtained in last step. Since the fitness space has multiple responses or objective values, therefore, it is required to obtain the Pareto fronts for the solutions.
Step 3. To obtain the Pareto fronts, a non-dominated sorting method is applied and each of the members of the population are ranked based on the domination.
Step 4. For every iteration of the heuristic, it generates a new population, evaluates the solutions, add them with initial population set, and rank them using the non-dominated sorting. Hence, each new population has now 200 members.
Step 5. Only the top 100 members are selected based on the ranks and crowding distances and passed on to the next iterations.

The heuristic stops, once it reaches the maximum number of iteration count.

4 Results and Discussions

The proposed algorithm is coded in Matlab. The obtained Pareto fronts for each of the test problems are displayed in Figs. 2, 3, and 4.

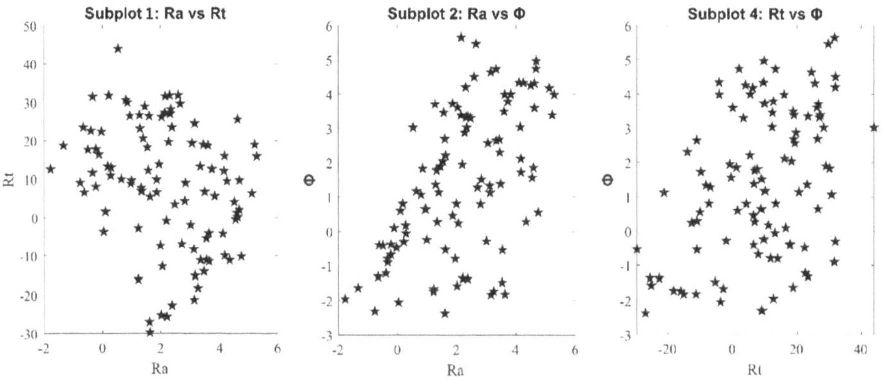

Fig. 2. Trade-off solutions for CNC turing case study.

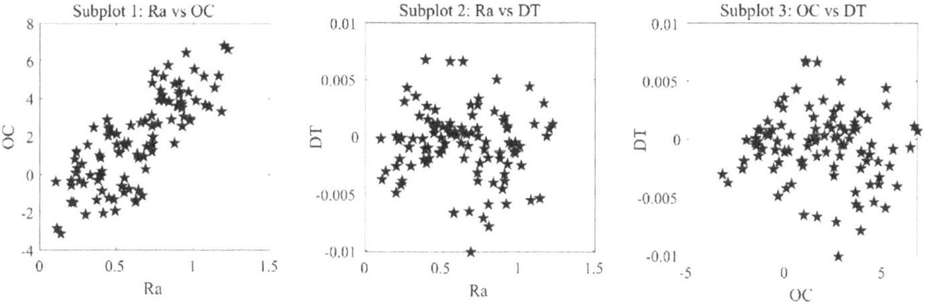

Fig. 3. Trade-off solutions for centerless cylindrical grinding case study.

The turning operation has three responses *Ra*, *Rt*, and φ; the grinding operation has three responses *Ra*, *OC*, and *DT*; and the milling operation has four responses, *TW*, *Fx*, *Fy*, and *Ra*. The computing time is an important factor for the optimization algorithms. This proposed technique consumes 12.14 s, 12.11 s, and 10.17 s to produce the optimal results, and the obtained results are better than the published results. For the CNC turning, the optimal published result is $CF = 155$, $FR = 0.12$, $DC = 0.8$, $CFMR = 12$, *and* $Ra = 0.484$, $Rt = 17.762$, $\varphi = 0.067$. In this study, the most suitably picked results are, *(1)* $CF = 131.08$, $FR = 0.14$, $DC = 0.9$, $CFMR = 13.47$, *and* $Ra = 0.269$, $Rt = 10.934$, $\varphi = 0.177$ *and (2)* $CF = 161.7$, $FR = 0.144$, $DC = 0.913$, $CFMR = 12.355$, *and* $Ra = 0.094$, $Rt = 1.552$, $\varphi = 0.604$. For the centerless cylindrical grinding, the optimal published result is the experimental run# 2 in Table 2. In this study two most promising solutions are picked up from the Pareto frontier, *(1)* $DF = 4.075$, $GF = 10.920$, $DT = 4.327$, $CT = 9.230$ *and* $Ra = 0.244$, $OC = 0.344$, $Dt = -0.004$ *and (2)* $DF = 4.260$, $GF = 5.959$, $DT = 2.894$, $CT = 9.958$ *and* $Ra = 0.229$, $OC = 0.830$, $Dt = -0.002$. For the micro milling, the optimal published result is

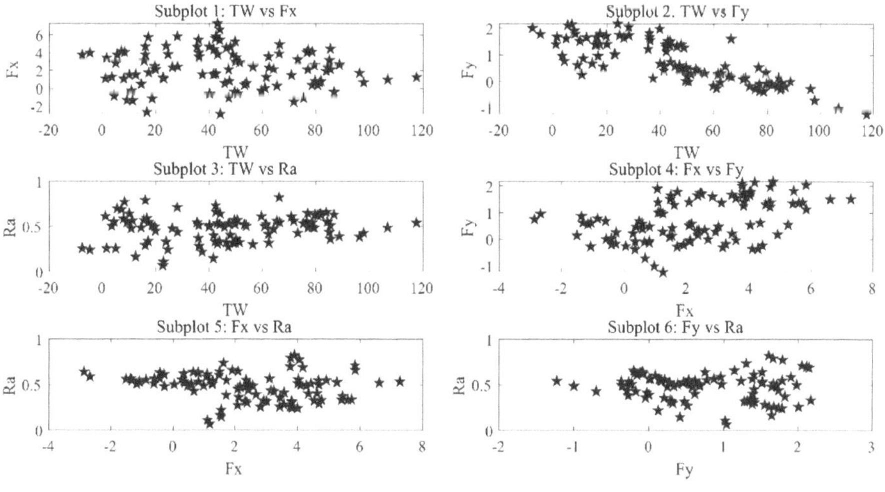

Fig. 4. Trade-off solutions for micro-milling case study.

the experimental run# 1 in Table 3. In this study, the most promising Pareto solutions obtained are, *(1)* $SS = 9373.596$, $FPT = 1.219$, $DC = 57.163$, $TW = 4.040$, $Fx = 2.799$, $Fy = 1.723$, $Ra = 0.198$ *and (2)* $SS = 11257.000$, $FPT = 0.153$, $DC = 86.430$, $TW = 4.494$, $Fx = 2.047$, $Fy = 1.863$, $Ra = 0.181$.

The surface roughness is the most exploited process response for all the cases, which determines the quality of the products. Therefore, the objective of this study is to find those Pareto solutions with the lowest surface roughness scores. Similarly, the Pareto solutions obtained using the proposed technique demonstrates better tool wear score for the CNC milling.

5 Conclusions

This paper proposes a generic surrogate-assisted heuristic algorithm for manufacturing process optimization problems. The proposed technique exploits the GKR surrogate, which is shown to outperform another surrogate based on the neural network. Both the surrogate functions need small amount of manufacturing process data for the training and obtain very accurate output responses with low NMSE scores. The proposed technique is successfully tested on three different cases from past literature. This study shows that the surrogate-assisted heuristic algorithm can obtain realistic solutions within boundary conditions. This study requires further validation tests for the reliability. The proposed surrogate-assisted approach could be extended further with standing optimization algorithms such as Nondominated Sorting Genetic Algorithm (NSGA), Multi-Objective Evolutionary Algorithm (MOEA) etc. This work is a work-in-progress of a project based on multi-response CNC milling operation and process monitoring.

Acknowledgement. This work is supported by the SFI Manufacturing (Project No. 237900) and funded by the Research Council of Norway (RCN).

References

1. Afazov, S.M.: Modelling and simulation of manufacturing process chains. CIRP J. Manuf. Sci. Technol. **6**(1), 70–77 (2013)
2. Gröger, C.F., Niedermann, F., Mitschang, B.: Data mining-driven manufacturing process optimization. In: Proceedings of the World Congress on Engineering. WCE, London (2012)
3. Mukherjee, I., Ray, P.K.: A review of optimization techniques in metal cutting processes. Comput. Ind. Eng. **50**(1–2), 15–34 (2006)
4. An, Y., Lu, W., Cheng, W.: Surrogate model application to the identification of optimal groundwater exploitation scheme based on regression kriging method - a case study of western Jilin Province. Int. J. Environ. Res. Public Health **12**(8), 8897–8918 (2015)
5. Stein, M.: Large sample properties of simulations using Latin hypercube sampling. Technometrics **29**(2), 143–151 (1987)
6. Chandrasekaran, M., Muralidhar, M., Krishna, C.M., Dixit, U.S.: Application of soft computing techniques in machining performance prediction and optimization: a literature review. Int. J. Adv. Manuf. Technol. **46**(5–8), 445–464 (2010)

7. Tzeng, C.J., Lin, Y.H., Yang, Y.K., Jeng, M.C.: Optimization of turning operations with multiple performance characteristics using the Taguchi method and Grey relational analysis. J. Mater. Process. Technol. **209**(6), 2753–2759 (2009)
8. Siddiquee, A.N., Khan, Z.A., Mallick, Z.: Grey relational analysis coupled with principal component analysis for optimisation design of the process parameters in in-feed centreless cylindrical grinding. Int. J. Adv. Manuf. Technol. **46**(9–12), 983–992 (2010)
9. Kuram, E., Ozcelik, B.: Multi-objective optimization using Taguchi based grey relational analysis for micro-milling of Al 7075 material with ball nose end mill. Measurement **46**(6), 1849–1864 (2013)
10. Baskar, N., Asokan, P., Prabhaharan, G., Saravanan, R.: Optimization of machining parameters for milling operations using non-conventional methods. Int. J. Adv. Manuf. Technol. **25**(11–12), 1078–1088 (2005)
11. Schetinin, V.: A learning algorithm for evolving cascade neural networks. Neural Process. Lett. **17**(1), 21–31 (2003)
12. Chapelle, O., Vapnik, V.: Model selection for support vector machines. In: Proceedings of the 12th International Conference on Neural Information Processing Systems, NIPS 1999, Denver, CO (2000)
13. Brentan, B., Laurain, V., Aberkane, S.: How to infer prior knowledge in water distribution data-driven models? In: WDSA/CCWI Joint Conference 2018, Kingston, Ontario, Canada (2018)

Assessment of Tool Wear Intensity Based on the Frequency Pattern

Anna Zawada-Tomkiewicz(✉) and Dariusz Tomkiewicz

Koszalin University of Koszalin, Koszalin, Poland
anna.zawada-tomkiewicz@tu.koszalin.pl

Abstract. The article is devoted to the issues of finishing machining, in which changes on the cutting edge significantly affect the quality of the machined surface. In tests EN 100Cr6 hardened steel has been used with a hardness of 60 HRC. The cutting tool material for turning tests were coated cemented carbides. Cutting parameters were selected so as not to exceed the permissible value of blade wear. The shortening of the tool has been evaluated from measurement of workpiece surface realized by stylus profilometer along the cutting length. This made it possible to obtain detailed information on blade wear over time. During cutting, the F_c, F_f and F_p components of the cutting forces were measured using a piezoelectric sensor and vibration signal was measured with a seismic sensor. The collected measurement data were then analyzed for diagnostic purposes using the Short Time Fourier Transform (STFT). First the stabilized process pattern was elaborated for one-second time windows. This pattern has been used to distinguish the excited states of the system including states of increased blade wear. Two wear intensity indicators have been determined for each of the recorded signals: mean deviation from the pattern and data dispersion. A moving average of wear intensity indicators has been applied for process state analysis.

Keywords: Machining · Diagnostics · Tool wear · Frequency pattern

1 Introduction

Hardened steel machining is analyzed in the context of finishing processes. For such processes, cutting is carried out by the rounded part of the cutting blade, which due to the properties of the workpiece is subjected to accelerated wear. Unlike the abrasive processes used, cutting of hardened steels requires constant supervision of tool wear [5, 9, 14, 15].

The wear of the tool affects the quality of the machined surface, because due to the kinematic and geometric mapping of the tool in the machined material, the shape of the corner of the blade imprints at feed intervals. Wearing the cutting blade is also related to its shortening, which causes the decrease of the thickness of the cutting layer (undeformed chip thickness) and affects the value of the workpiece dimension.

Tool wear is also considered in the context of intensity of tool geometry changes [2]. The wear intensity is defined as a derivative of tool wear index during cutting. Increased tool wear intensity influences both the low-frequency surface properties

M. Majewski and W. Kacalak (Eds.): IIRTS 2019, LNME, pp. 36–47, 2020.
https://doi.org/10.1007/978-3-030-37566-9_4

associated with the change in workpiece dimension (due to the shortening of the blade) and the high-frequency properties related to variable chip thickness (due to the blunting and jaggedness of the cutting edge) [11, 13, 17]. Changes in surface texture properties affect the dynamics of the cutting process, which can be registered in the signal of cutting force and vibrations [12, 16].

The purpose of the research was to analyze the components of the cutting force and vibrations during cutting of hardened steel under the finishing conditions. Research methodology is presented in Sect. 2. Due to the fact that non-stationary signals were analyzed, Short Time Fourier Transform was applied. It allowed frequency analysis of cutting force components and vibration signals during cutting. The result of the transformation has been analyzed in time windows and for such images, collected in time, the comparisons have been made with reference images. The process model and signal processing method are presented in Sect. 3. Section 4 is a summary of the research.

2 Research Methodology

The purpose of the research was to analyze the components of the cutting force and vibrations during cutting of hardened steel under the finishing conditions. TNMG 160408 PF 4015 blades from Sandvik were used in turning tests. The view of the cutting insert with the measurement of the cutting edge rounding radius is given in Fig. 1a. The tests were carried out on rods made of EN 100Cr6 bearing steel with a hardness of 60 HRC. Cutting parameters were selected to be the following: feed rate 0.15 mm/rev, cutting speed 151 m/min, depth of cut 0.15 mm. The photographs of tool flank wear after 15 s of cutting is presented in Fig. 1b. The *VB* wear index value (maximum wear width of the minor flank) has been estimated to be 0.56 mm.

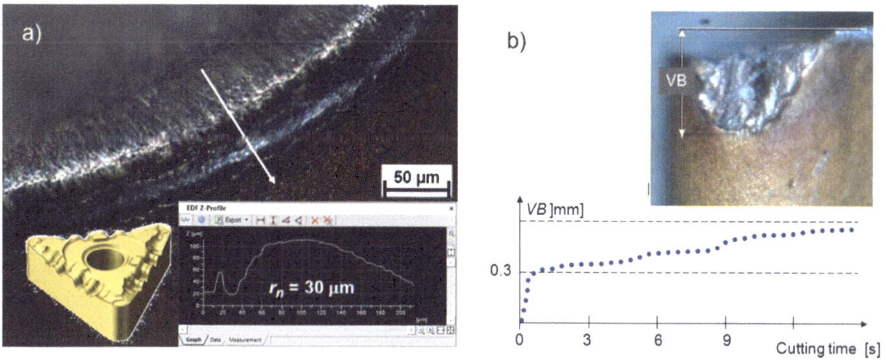

Fig. 1. Methodology of research: (a) cutting tool and cutting edge radius measurement before the cutting process using a Nikon MA200 microscope, (b) the photograph of tool flank wear and chart of *VB* tool flank wear index over cutting time (description in the text).

For tests of the blade wear process over time, the value of the *VB* wear index has been correlated with the shortening of the blade measured with a profilometer. For 2D measurements, the Talysurf CLI 2000 with a contact head was used. Thanks to this methodology, the values of the *VB* wear index over time were obtained. It has been possible to determine the intensity of the change of wear over time.

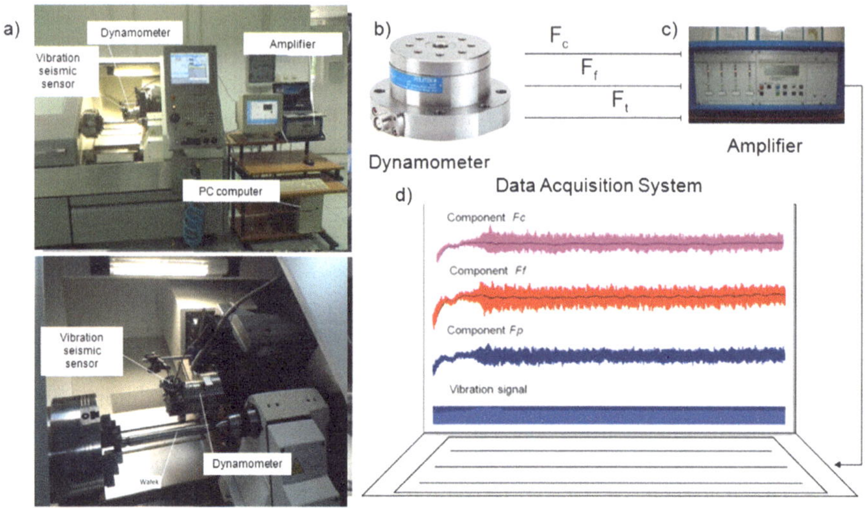

Fig. 2. Cutting force and vibration measurement system: (a) NEF 400 machine tool, (b) dynamometer Kistler 9272, (c) amplifier Kistler 5019, (d) DAQ LabView software.

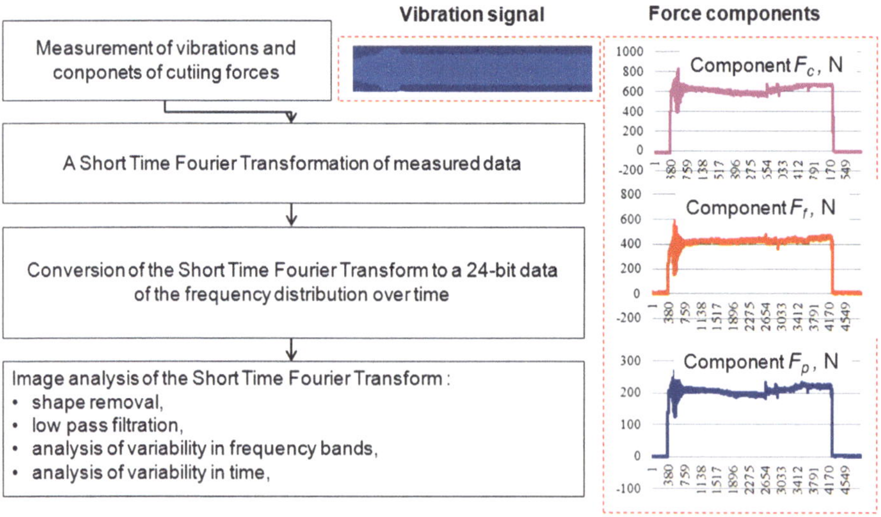

Fig. 3. Methodology for analyzing cutting force components and vibration signals.

The cutting process was tested on a NEF 400 lathe (Fig. 2). A Kistler 9272 piezoelectric dynamometer was mounted in the VDI 30 tool chuck, which allowed the measurement of the values of force components: peripheral F_c, feed F_f and resistance F_p. A seismic sensor with 70 dB sensitivity and 20,000× sound amplification was also attached to the tool holder.

Methodology for analyzing cutting force components and vibration signals is shown in Fig. 3. The measurement data were transformed using Short Time Fourier Transformation. The transform as a result of the transformation was presented in later chapters (Fig. 4, 5, 6, 7 and 8) as a colored amplitude map, where time is on the horizontal axis and frequency is on the vertical axis. The maps were analyzed by image processing methods.

3 Model of the Cutting Process

If it can be assumed that the measured data are stationary, then methods like for stationary signals can be used. A stationary random signal is a signal for which all statistical moments are unchanged in time. In such cases the spectral analysis can determine the power distribution as a function of frequency for specific time moments [1]. Analyzing the machining process, there can be distinguished states in which the process is carried out in a stable manner, and states in which this stability is lost. Analyzing the measured signals over time for different frequency bands, it was found that the entire frequency range should be analyzed assuming a known time. Just like in case of rapid wear of the tool where the force and vibration signals change significantly [4, 6, 8]. In this case, data analysis must be carried out by means of time-frequency analysis as for non-stationary signal.

Short Time Fourier Transform (STFT) is a solution to the problem of signal spectrum variation. The Fourier transform is calculated in a fixed-length window that moves on the time axis so that a series of time-ordered spectra is obtained. This allows you to observe changes in the spectrum over time [10].

The Short Time Fourier transform is a compromise between the representation of the signal in time form and frequency. It provides information on both of these parameters, i.e. the frequencies occurring in the tested waveform, as well as information when a given frequency appears in the signal. However, this information can be read with quite limited accuracy, which is affected by the time length of the window used (to facilitate this task, the charts are two-dimensional). Once assumed, the window length is the same for all spectral components of the waveform, which is not quite flexible due to the variable nature of the frequency components appearing in the considered signal. The disadvantage of this approach is the limited resolution in the frequency domain [3, 13].

Figure 4 presents a chart of the Short Time Fourier Transform of the vibration signal. Two frequency bands A and B are marked on the map, and then they are drawn as a function of time – amplitude. The amplitude changes over time and stabilizes within 2 s. During approx. 9.5 s the amplitude has increased. This increase is clearly visible in the map. Nevertheless, in the A and B charts the course of the amplitude of vibrations for these bands is noticeable but distributed in time. As can be seen from the compiled amplitude data for the A and B bands, it should be noted that not only the value of the amplitude in time is important for wear intensity diagnostics but also its dispersion can be used as diagnostic measure.

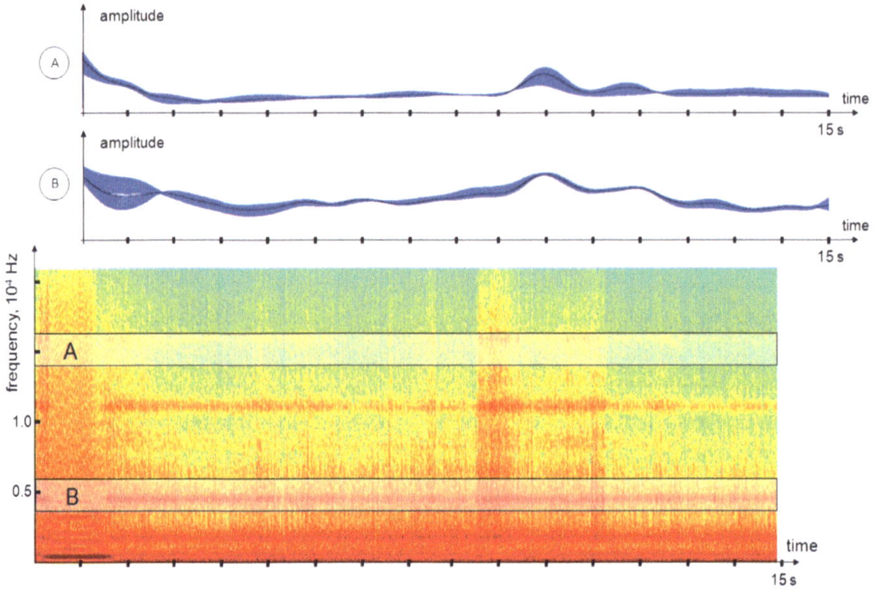

Fig. 4. STFT of the vibration signal and its analysis in time.

When analyzing the machining process as a function of technological factor and stability of the cutting process, several different process states can be distinguished. Start of the cutting process initiates the increase of the cutting force, which after a short time stabilizes on the level being the result of the process parameters. Dynamical part of the cutting force for stabilized process is set on a given level. Similar behavior is observed in the vibration signal.

The occurrence of interference in the machining process can lead to a loss of stability [6]. Initially, the machining system falls into vibrations, which gradually intensify, making the dynamic system unstable. Statistical tests for vibration signals and cutting force for blades with a high degree of wear and during the tool wearing process allowed distinguishing two additional states in which the cutting process may be found.

States of the cutting process with an example of a vibration signal were presented in Table 1. Portion of the STFT data corresponding to 1 s time window were filtered (form removal and low-pass filtration) and an envelope of amplitudes was created in frequency function. Such data were treated as a pattern of a cutting process state. Samples of the vibration signal and cutting force components for determining the pattern have been obtained in the same cutting conditions but in different cutting tests.

Similar results of the stabilized process pattern and the wear intensity pattern have been obtained for cutting force components. The determined patterns have been used to analyze the cutting process. Based on the patterns, two wear intensity indicators were created: the average deviation from the stabilized process pattern (average distance between measured data set and the stabilized process pattern) and data dispersion (sum of square errors of absolute difference between measured data set and the stabilized process pattern). These indicators were normalized when used in diagnostics.

Table 1. States of the cutting process with an example of a vibration signal.

State of the process	STFT of the signal (1 s)	STFT of the signal after filtration	Pattern
Stable process pattern	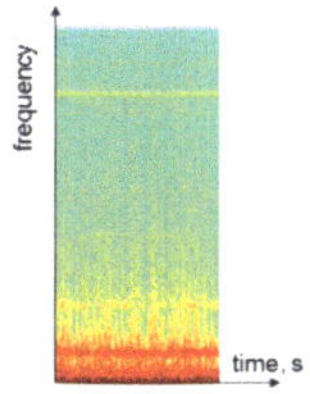	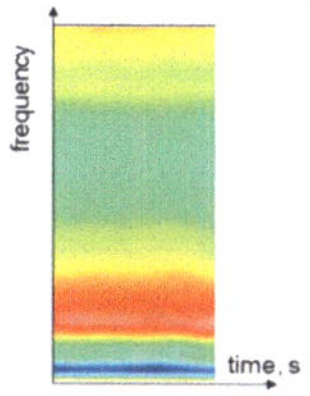	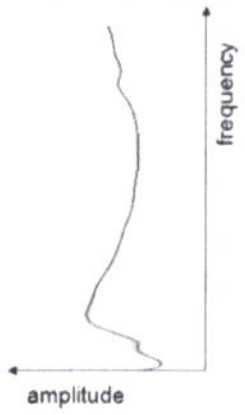
The stabilized process with defined production capacity (A) is defined by a technological process, there are relationships between samples of given signals coming from the machining system.	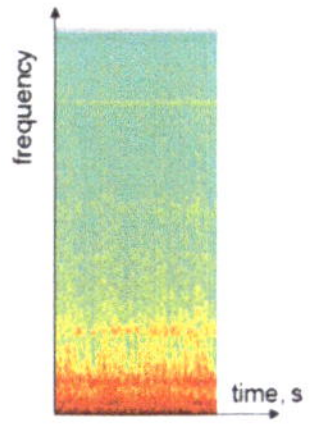	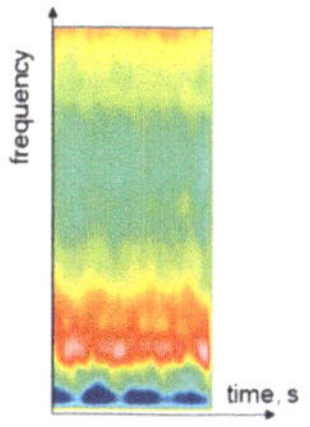	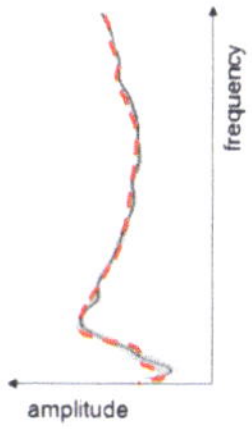
The process of deterministic chaos (B), which is deterministic, small changes in process parameters can cause significant changes in the location of the process in phase space, and the value of the largest Lapunov coefficient is greater than zero.	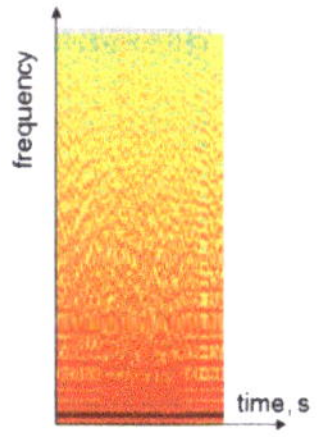	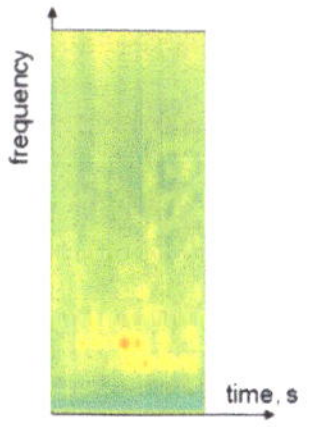	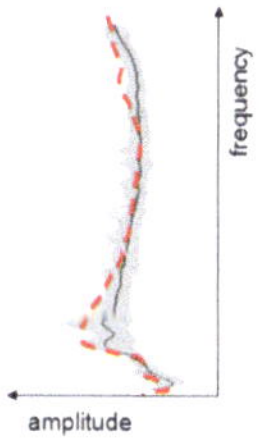
The unstable process (C), in which signals from the machining system increase periodically, and the value of the largest Lapunov coefficient is zero.	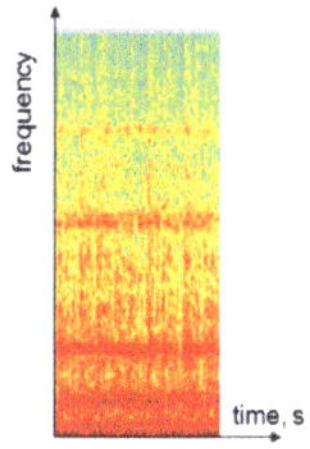	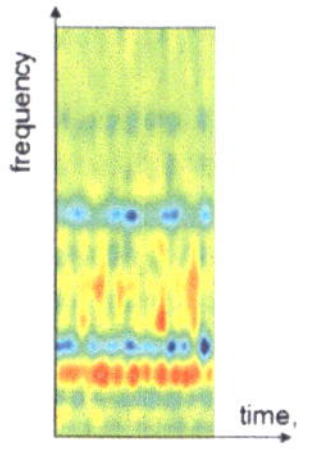	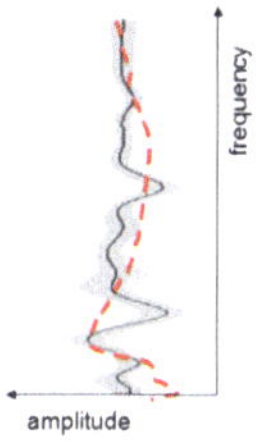

The STFT of the whole vibration signal was calculated. Portions corresponding to 1 s were taken from these data, which were then filtered. Form removal and low pass filtration were applied. The data set obtained in this way was converted into image and the pattern was calculated as the average frequency value over 1 s.

As can be seen in Fig. 5, for selected portions of data there are significant differences between the stable process pattern and the one obtained in the cutting process. Similar results of the STFT analysis for individual components of the cutting force are shown in Fig. 6, 7 and 8.

Figure 6 shows the STFT for the cutting force component F_c. As can be seen, apart from the initial period of increased wear intensity, there are also areas in the range of 9–10 s and 12.5–13.5 s. In these areas there are significant differences between the stabilized process pattern and the pattern for increased intensity.

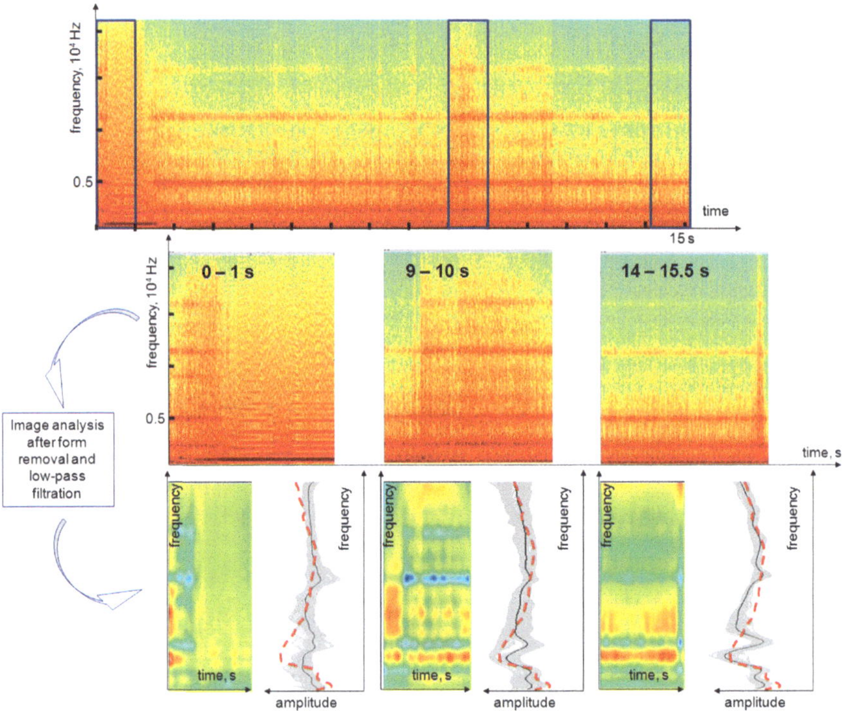

Fig. 5. STFT of the vibration signal with an example of analysis and comparison to the pattern of one-second portions of data.

In the case of the F_f component of the cutting force, the process intensity and a significant deviation from the pattern are visible only for the initial period of high wear intensity and in the range of 14–15.5 s cutting (Fig. 7). In contrast, the STFT of the F_p component of the cutting force revealed an initial period of intense wear that significantly changes the cutting edge (Fig. 8).

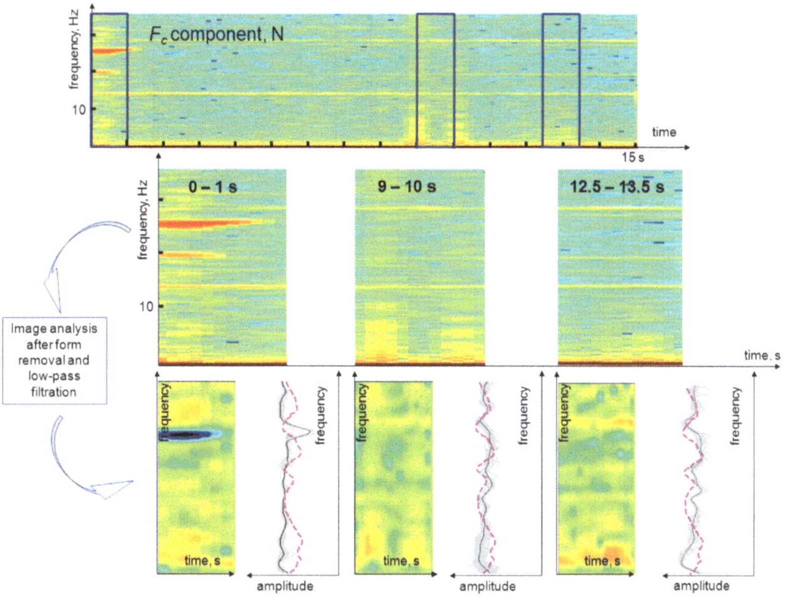

Fig. 6. The STFT of the force component F_c with an example of analysis and comparison to the pattern of one-second portions of data.

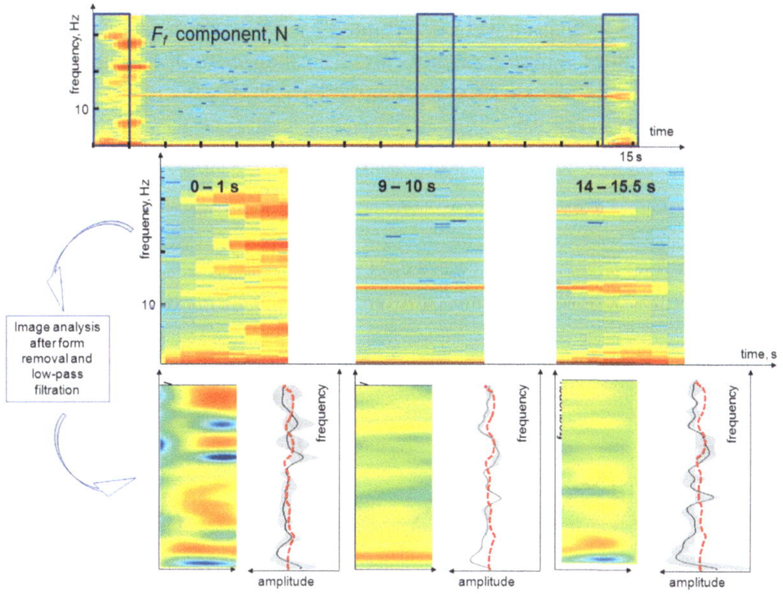

Fig. 7. The STFT of the force component F_f with an example of analysis and comparison to the pattern of one-second portions of data.

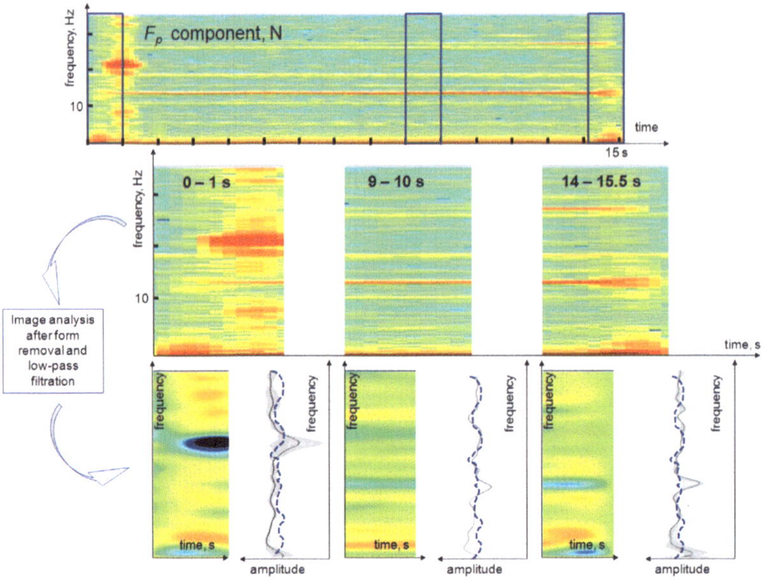

Fig. 8. The STFT of the force component F_p with an example of analysis and comparison to the pattern of one-second portions of data.

Data portions corresponding to 1 s were analyzed for the whole cutting time. Two previously mentioned indicators were determined for them: deviation from the pattern and dispersion of pattern. These indicators were normalized to the range [0, 1]. Next, the moving average of the indices was compiled so as to obtain 465 points, which corresponded to 30 Hz (Fig. 9).

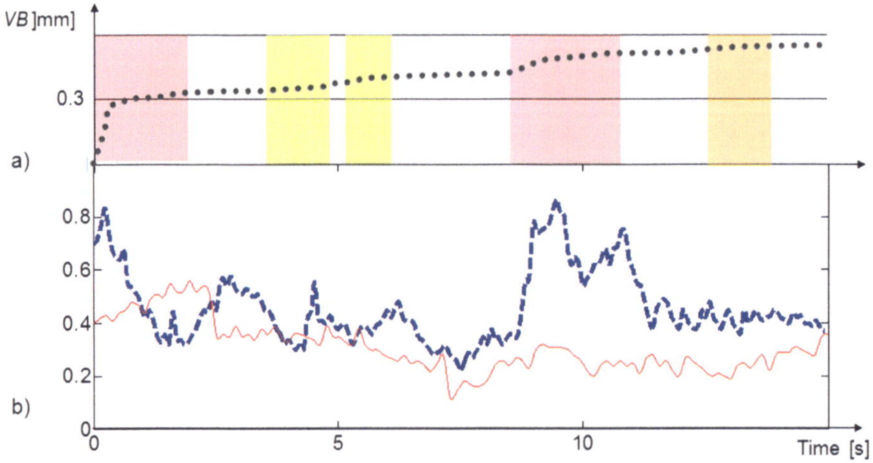

Fig. 9. The normalized value of the deviation between the measured value and the pattern of the stabilized process (dashed line) and the normalized value of data dispersion (solid line).

The normalized value of the deviation between the measured value and the pattern of the stabilized process (dashed line) and the normalized value of data dispersion (solid line) is shown in Fig. 5b. Five different colored boxes are marked against the background of tool wear. Red color of field has been selected for designation of high intensity of wear, orange for lower and yellow for the smallest intensity of wear. The visualized course of tool wear intensity indicators for the vibration signal shows that they are able to indicate the states marked as red, with high tool wear intensity. Lower intensities of tool wear are masked by system disturbances. Similar results were obtained for cutting force components.

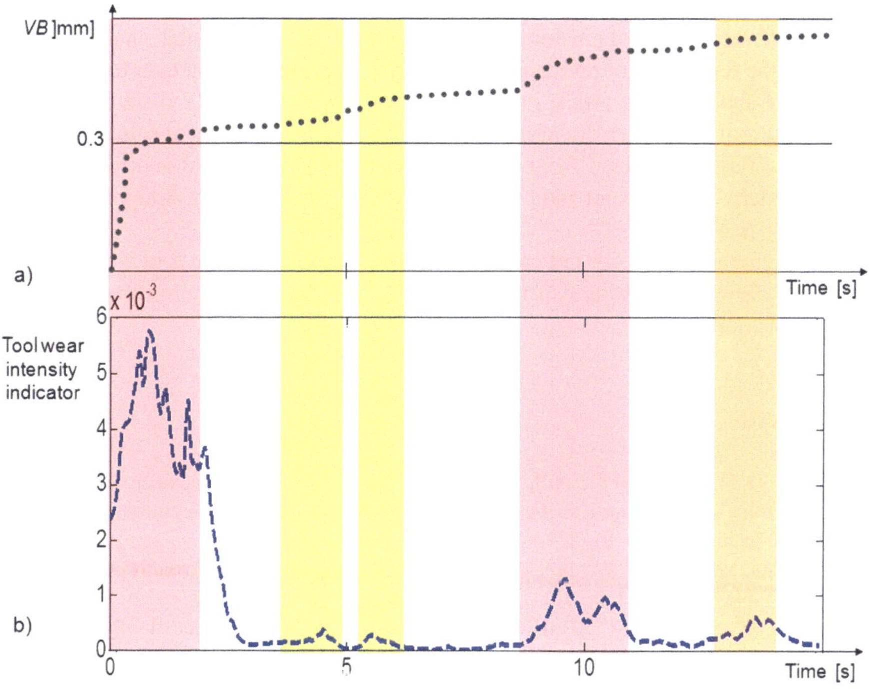

Fig. 10. (a) Tool flank wear and (b) tool wear intensity indicator vs. time.

As mentioned earlier, individual results of vibration signal intensity indicators and cutting force components were not able to detect changes occurring in the cutting process. This is in line with the results described in [4] and [7]. The tool wear intensity indicator referring to the deviation of the measured data from the pattern was able to detect each time a high intensity of tool wear, while the lower intensity of wear was masked by interference. In the case of the indicator referring to the dispersion of measurement data within 1 s, in the case of the vibration signal it was masked with interference, while in the case of cutting force components it was able to detect a higher intensity of blade wear.

Figure 10 shows the aggregated intensity index for eight partial indicators (2 vibration indicators, 6 cutting force components). Visualization of the aggregated intensity indicator of tool wear denotes its high sensitivity not only to the high intensity of wear but also to medium and even low wear intensity.

4 Conclusions

The work presents a time-frequency analysis of the vibration signal and the signal of cutting force components during turning of hardened steel with a coated cemented carbide tool, which was subjected to intensive wear during cutting.

The recorded vibration signals and cutting force components were analyzed using the Short Time Fourier Transform, which enabled not only visual analysis of the process in the form of a colored map showing frequency distribution over time, but also quantitative analysis. The article proposes two tool wear intensity indicators. The first indicator of tool wear intensity referrers to the determination of the absolute difference between data set determined from the STFT of the analyzed signal and the stabilized process pattern. The second indicator of tool wear intensity was designated as frequency dispersion for 1 s.

The aggregated intensity indicator developed from eight partial tool wear intensity indicators allowed to distinguish three levels of process intensity resulting from the intensity of tool wear.

References

1. Ahmed, Y.S., Alam, M.S., Arif, A.F.M., Veldhuis, S.C.: Use of acoustic emission and cutting force signals to monitor built-up edge formation in stainless steel turning. Int. J. Adv. Manuf. Technol. **103**(5–8), 2257–2276 (2019)
2. Bartoszuk, M., Grzesik, W.: Investigation of initial wear period of differently coated carbide cutting tools. J. Mach. Eng. **15**(4), 37–45 (2015)
3. Chen, G.X., Zhou, Z.R.: Time–frequency analysis of friction-induced vibration under reciprocating sliding conditions. Wear **262**, 1–10 (2007)
4. Duro, J.A., Padget, J.A.: Bowen, C.R., Kim, H.A., Nassehi, A.: Multi-sensor data fusion framework for CNC machining monitoring. Mech. Syst. Sig. Process. **66–67**, 505–520 (2016)
5. Kacalak, W., Lipiński, D., Bałasz, B., Rypina, Ł., Tandecka, K., Szafraniec, F.: Performance evaluation of the grinding wheel with aggregates of grains in grinding of Ti-6Al-4V titanium alloy. Int. J. Adv. Manuf. Technol. **94**, 301–314 (2018)
6. Kalvoda, T., Hwang, Y.-R.: Analysis of signals for monitoring of nonlinear and non-stationary machining processes. Sens. Actuators A **161**, 39–45 (2010)
7. Kannatey-Asibu, E., Yum, J., Kim, T.H.: Monitoring tool wear using classifier fusion. Mech. Syst. Sig. Process. **85**, 651–661 (2017)
8. Lauro, C.H., Brandão, L.C., Baldo, D., Reis, R.A., Davim, J.P.: Monitoring and processing signal applied in machining processes - a review. Meas. J. Int. Meas. Confed. **58**, 73–86 (2014)
9. Lipiński, D., Kacalak, W.: Metrological aspects of abrasive tool active surface topography evaluation. Metrol. Meas. Syst. **23**(4), 567–577 (2016)

10. Song, K.I., Cho, G.C.: Numerical study on the evaluation of tunnel shotcrete using the Impact - Echo method coupled with Fourier transform and short - time Fourier transform. Int. J. Rock Mech. Min. Sci. **47**, 1274–1288 (2010)
11. Storch, B., Zawada-Tomkiewicz, A.: Distribution of unit forces on the tool nose rounding in the case of constrained turning. Int. J. Mach. Tools Manuf. **57**, 1–9 (2012)
12. Wang, G.F., Yang, Y.W., Zhang, Y.C., Xie, Q.L.: Vibration sensor based tool condition monitoring using ν support vector machine and locality preserving projection. Sens. Actuators A: Phys. **209**, 24–32 (2014)
13. Zawada-Tomkiewicz, A.: Machined surface quality estimation based on wavelet packets parameters of the surface image. PAK **6**, 606–609 (2010)
14. Zawada-Tomkiewicz, A., Tomkiewicz, D.: Mechatronic monitoring tool (MMT) for machined surface evaluation during the cutting process. In: Proceedings of 11th IMEKO TC14 Symposium on Laser Metrology for Precision Measurement and Inspection in Industry, LMPMI, pp. 373–376 (2014)
15. Zawada-Tomkiewicz, A., Ściegienka, R., Zuperl, U., Stępień, K.: Images of the machined surface in evaluation of the efficiency of a micro-smoothing process. Strojniški vestnik – J. Mech. Eng. **65**, 7–8 (2019)
16. Zawada-Tomkiewicz, A., Storch, B.: Analysis of components of the cutting force during turning of hardened steel under conditions of finishing treatment. Mechanik **8–9**(725), 381–389 (2015)
17. Zawada-Tomkiewicz, A., Storch, B.: Classifying the wear of turning tools with neural networks. J. Mater. Process. Technol. **109**, 300–304 (2001)

Statistical Process Control Accuracy Estimation of a Stamping Process in Automotive Industry

Radu Godina[1(✉)] and João C. O. Matias[2,3(✉)]

[1] Research and Development Unit in Mechanical and Industrial Engineering (UNIDEMI), Faculty of Science and Technology (FCT), Department of Mechanical and Industrial Engineering, Universidade NOVA de Lisboa, 2829-516 Caparica, Portugal
r.godina@fct.unl.pt

[2] DEGEIT, University of Aveiro, 3810-193 Aveiro, Portugal
jmatias@ua.pt

[3] GOVCOPP, University of Aveiro, 3810-193 Aveiro, Portugal

Abstract. The goal of this paper is to unravel a quality problem, mainly the quality improvement of a process by employing statistical methods. The normality test utilized in the case is cumulative frequency distribution with regression analysis. The statistical process control (SPC) technique control charts reveal that the process is centered and meets the acceptance criteria and the regression analysis reveals that the recorded data follow a normal distribution. In this paper the results of the K-S, Anderson-Darling and Shapiro-Wilk tests are analyzed and discussed since these tests are, according to the literature more powerful statistical tools for detecting most departures from normality. In order to estimate more accurately if the tested data came from a normally distributed population, three goodness of fit tests were performed on the same collected values: Kolmogorov-Smirnov, Anderson-Darling and Shapiro-Wilk. Then, the results were analyzed, discussed and compared. Since the tests are more sensible to detect most departures from normality, they allow a more accurate assessment of the collected data – in turn increases the confidence in the control chart.

Keywords: Control chart · Quality control · Process capability index · Stamping process · Statistical process control · Automotive industry

1 Introduction

A fundamental part of any organization's business strategy is the determination of its competitive priorities, which must be aligned with its vision of the future and which represent its sustainability cards in the market [1]. Some years ago, it was common for organizations to discuss which objectives would reinforce their business strategies, whether it would be cost, quality, service, flexibility or innovation. However, nowadays, quality is not optional if it is to survive as a company in a globalized environment, it became an intrinsic factor of any competitive organization, and failure to comply with its specifications is the first step towards leaving the market [2].

M. Majewski and W. Kacalak (Eds.): IIRTS 2019, LNME, pp. 48–60, 2020.
https://doi.org/10.1007/978-3-030-37566-9_5

The automotive industry is facing an increasing demand and uncertainty due to the growing market competition, shortened product life cycles, and high product differentiation. A major challenge for manufacturing companies is to offer quality products without increasing production costs and the most basic concern in an organization cantered around quality is to what degree is possible to satisfy the clients' requirements. If such requirements are established, it is essential to quantify how the organization is able to satisfy the clients' expectations [3]. With the intent to cope with the several requirements of their clients, enterprises idealize and develop products with characteristics that satisfy such requirements. Thus, processes that can match the characteristics of such requirements must be implemented as well as adequate control and monitoring [2].

The quality control is a set of tools that are integrated into a methodological process to incorporate quality in all phases of development including product design, manufacturing and process control. The professional in charge of the quality control process performs or participates in the characterization of the new products or services in their different phases of development and in the establishment of their quality specifications. In the same way that develops, executes or coordinates the execution of the test methods to determine the quality characteristics of raw materials, materials, intermediate products and final products. For this purpose, statistical tools are used for the product design and process quality improvement [4].

The statistical thinking relies on principles that all work takes place in interrelated processes. Variation occurs in all processes thus reducing it is the cornerstone of the process improvement. A practical utilization of statistical techniques such as SPC obliges to an established implementation framework by statistical thinking [5]. Industrial enterprises are trying more to improve the production performance concerning the productivity output with the intention of remaining competitive due to the reason that a great productivity performance has a direct relationship with the process control and equipment efficiency [6]. Lack of suitable productivity measurement indicators means for the industrial enterprise an unidentified performance. An unstable process that is not under control can result in manufacturing of nonconforming products, therefore affecting the general production performance [7].

The SPC is an integrated assembly of problem-solving tools created with the intention to reach the stability of the manufacturing process and increase the capability by decreasing the variability [8]. The SPC could be employed when great quantities of the same item are being manufactured [9]. The fundamental basis on which SPC relies is that conforming items are produced when the process is deemed in control by satisfying the client requirements [10]. The key goal of SPC is to highlight a signal when the process alters, that is to say when the mean drifts away from the target and/or its variability grows [11]. As soon as a signal indicates that the process is altering, the machine operator has to initiate a corrective operation. By using this method, products of the ideal quality will be not manufactured due to visual quality inspection but rather because they are manufactured properly, which is less costly. Usually, the key SPC tool is the control chart, which represents in detail the process data and indicates whether the process is under statistical control or not [12].

In contemporary service and industrial enterprises, SPC control charts have been widely employed [13]. At the beginning for the quality control Shewhart advocated for the utilization of SPC control charts for monitoring the process variation in 1924. The process variation behaves in a certain nature: exceptional causes lead to atypical changes, while non-exceptional causes produce normal changes [14]. In this perspective, a process is deemed to be "in statistical control" if the probability distribution defining the quality characteristic remains unchanged during the process operation. If some change occurs in this distribution, the process is known as "out of control" [15].

The goal of this paper is to unravel a quality problem, mainly the quality improvement of a process by employing statistical methods. In this paper an examination of a case study of an automotive SME located in Portugal where SPC is applied is performed. The normality test utilized in this case is cumulative frequency distribution with regression analysis. The SPC technique control charts reveal that the process is centred and meets the acceptance criteria and the regression analysis reveals that the recorded data follow a normal distribution. Nonetheless, since the Shapiro-Wilk, K-S and Anderson-Darling tests are more powerful than the one currently in use, when the simple regression analysis test is substituted by the aforementioned tests the results could alter. In this paper the results of the K-S, Anderson-Darling and Shapiro-Wilk tests are analyzed and discussed since these tests are, according to the literature [16], more powerful statistical tools for detecting most departures from normality. The novelty lays in the fact that in the literature little or no publications research such approach since the normal distribution is always assumed to be normal which in this study the authors argue that is not always the case.

This organization of this paper is as follows. In Sect. 2 the alternative goodness of fit tests is succinctly addressed. The depiction of the targeted industrial unit and the analysis of the process, data collection and control charts are shown in Sect. 3. In Sect. 4 an analysis of the acquired results and their discussion is effectuated. Finally, the conclusion remarks are made in Sect. 5.

2 The Process Capability Index and Normality Tests

2.1 The Process Capability Index

In the attempts usually made for the process improvement, process capability ratio or the process capability index (C_{pk}) is known as a statistical measure of process capability: the capacity of a process to deliver a certain output within the limits of several previous stipulated specifications. The notion of process capability is only relevant in cases of processes that are in a state of statistical control [9]. The process capability indices quantify how ample the natural variation a process is when compared to its specification limits. Such types of indices also allow different processes to be compared regarding how well the manufacturing unit controls them [9]. The founding index to be published in the literature was the process precision index C_p and is expressed as:

$$C_p = \frac{L_{us} - L_{ls}}{6\sigma} \tag{1}$$

where L_{us} and L_{ls} are the lower and upper specification limits of the process, respectively. The estimated mean of the process is expressed by μ and the standard deviation of the process, also known as estimated variability, is represented by σ. However, the most typically recognised process capability indices comprise the following both:

$$C_p^l = \frac{\mu - L_{ls}}{3\sigma} \tag{2}$$

and:

$$C_p^u = \frac{L_{us} - \mu}{3\sigma} \tag{3}$$

The index C_p was created with the purpose to measure the extent of the overall process variation concerning a given manufacturing tolerance. Since the entire concept of this index is quite simple, C_p is not able to reveal the direction of process centring and, as a consequence, it does not give any indication of the actual process performance. Therefore, the C_{pk} index was created by taking into consideration essential details such as the process location as well as magnitude of the process variation. The C_{pk} index is defined by the following expression:

$$C_{pk} = min\left\{\frac{L_{us} - \mu}{3\sigma}, \frac{\mu - L_{ls}}{3\sigma}\right\} \tag{4}$$

Thus, the equation can be rewritten as:

$$C_{pk} = \delta - |\mu - \zeta| \tag{5}$$

where μ represents the process mean of the quality attribute, δ represents the half-length of the specification interval:

$$\delta = \frac{L_{us} - L_{ls}}{2} \tag{6}$$

and ζ is the midpoint of the specification interval:

$$\zeta = \frac{L_{us} + L_{ls}}{2} \tag{7}$$

The quantification of the process capability indices is represented through a dimensionless number, where higher the value higher the capability of the process will be. When the numbers are near or below zero means that the processes are operating off target or with high variation. Setting limits for the minimum approvable process capability targets has to be determined beforehand. Such limits vary depending on the nature of the facility, industry, and the process in question.

In case of a C_{pk} at level 1, in terms of statistics, it would be expected that the fabrication of nonconforming pieces is as a maximum 2 700 parts per million (PPM), which means outgo outside the specification limits. However, at a C_{pk} level of 1.33, the defect rate descends to 66 PPM.

The minimum C_{pk} value is usually specified in the acquiring contract, before the project even begins. If defining the minimum C_{pk} is unsuccessful, the process is considered to be incapable. However, if defining the minimum is successful the process is defined as capable. Montgomery [9] argues that several minimum capability requirements are recommended for processes runs under certain designated quality conditions, especially $C_{pk} \geq 1.33$ for existing processes.

2.2 The Kolmogorov-Smirnov Test

The Kolmogorov-Smirnov Test (K-S) compares the collected data with a known distribution and indicates if they have the same distribution. The test is nonparametric, meaning that it does not assume any particular underlying distribution—it is commonly used to assess if the data is normally distributed. K-S has two versions: one-sample test and two-sample test. The K-S statistic is used to quantify a distance between the cumulative distribution function of a reference distribution and the empirical distribution function of the sample, or among the empirical distribution functions of two samples. The null distribution in the case of one-sample of such type of statistical test is assessed via the null hypothesis H_0 that the sample is drawn from the reference distribution. The distributions taken into consideration under the H_0 are continuous distributions or else unrestricted.

Assuming b_i are n independent and identically distributed observations of a continuous value then the empirical cumulative distribution function CDF can be expressed as:

$$CDF = \frac{1}{n}\sum_{i=1}^{n} I^{f}_{[-\infty,\, b]}(b_i) \tag{8}$$

where I^f represents the indicator function. I^f is 1 if $b_i \leq b$, otherwise is 0.

By being $x_1, \ldots, x_n$ an ordered sample with $x_1 \leq \ldots \leq x_n$ then $K_n(x)$ can be represented as the following:

$$K_n(x) = \begin{cases} 0, & x < x_1 \\ k/n, & x_k \leq x < x_{k+1} \\ 1, & x > x_n \end{cases} \tag{9}$$

By assuming that the sample is originated from a population with a cumulative distribution function $CDF(x)$, then D_n can be defined as:

$$D_n = \max_x |CDF(x) - K_n(x)| \tag{10}$$

The purpose becomes to utilize D_n as a method that allows to assess the $CDF(x)$. Therefore, the distribution of D_n is assessed as the following:

$$CDF(x) = \frac{\sqrt{2\pi}}{x}\sum\nolimits_{k=1}^{\infty} e^{-(2k-1)^2\pi^2/8x^2} \tag{11}$$

Yet for the aims of this study the significant trait of the distribution are the critical values - $D_{n,\alpha}$ taken from Kolmogorov distribution table [17]. By taking $D_{n,\alpha}$ from the table, then it is applied:

$$P\left(D_n \leq D_{n,\alpha}\right) = 1-\alpha \tag{12}$$

can be utilized to test the hypothesis of a random sample being from a population with a specific distribution function $CDF(x)$. Thus, if:

$$\max_x |CDF(x) - K_n(x)| \leq D_{n,\alpha} \tag{13}$$

is verified then the sample data is considered a good fit with $CDF(x)$.

2.3 The Shapiro-Wilk Test

The test coined by Shapiro-Wilk utilizes the null hypothesis postulate to attest whether the sample $b_1, \ldots, b_n$ is created from a normally distributed population or not. Thus, if the p-value is greater than 0.05, then the null hypothesis is not rejected and thus there is indication that the tested data has its origins in a normally distributed population. The Shapiro-Wilk statistic S_W relies on the ratio of two estimates of scale [18] and is given by:

$$S_W = \frac{\left(\sum_{i=1}^{n} a_i b_{(i)}\right)^2}{\sum_{i=1}^{n} (b_i - \bar{b})^2} \tag{14}$$

where $b_1, \ldots, b_n$ represents a sample of continuous distribution and $b_{(1)}, \ldots, b_{(n)}$ represents the order statistics of $b_1, \ldots, b_n$, a_i embodies a constant and in which $b_{(i)}$ is the ith order statistic, i.e., the ith lowest number in the sample. The sample mean $\bar{b}$ is given by:

$$\bar{b} = \frac{(b_1 + \ldots + b_n)}{n} \tag{15}$$

The constant a_i is given by:

$$a_i = (a_1, \ldots, a_n) = \frac{k^{\mathrm{T}} V^{-1}}{\left(k^{\mathrm{T}} V^{-1} V^{-1} k\right)^{1/2}} \tag{16}$$

where:

$$k_i = (k_1, \ldots, k_n)^{\mathrm{T}} \tag{17}$$

where $k_1, \ldots, k_n$ are the estimated values of the order statistics of independent and equally distributed random variables tested from the standard normal distribution and V gives the covariance matrix of the above-mentioned statistic order.

2.4 The Anderson-Darling Test

By attributing a weight to both tails of the distribution, the Anderson-Darling test is performed with the purpose to increase the bias at the tails of a distribution. Also, for the purpose of test statistics, this test offers specific equations with the intent to assess the disparity between two distribution functions [19].

The statistical test is utilized to assess whether a given sample of a given array of data is taken from a certain probability distribution. In its elementary shape, the test makes an assumption that in the distribution to be tested no parameters exist to be estimated, meaning that the test and its set of vital values is distribution-free [20].

In statistics, when the cumulative distribution function is interrelated with the empirical measurement of the sample is known as the empirical distribution function (EDF). Anderson and Darling proposed a different type of quadratic EDF test represented by the following equation:

$$P_n = n \int_{-\infty}^{\infty} [J_n(x) - J(x)]^2 \lambda(x) dJ(x) \tag{18}$$

where $\lambda(x)$ represents the weight function, the cumulative distribution function is represented by J, and J_n represents the empirical distribution function. While the following equation is obtained when the weight function has the following behaviour:

$$\lambda(x) = [J(x)\{1 - J(x)\}]^{-1} \tag{19}$$

which modifies the Anderson-Darling test, A_n^2 as follows:

$$A_n^2 = n \frac{[J_n(x) - J(x)]^2}{J(x)\{1 - J(x)\}} dJ(x) \tag{20}$$

The Anderson-Darling test weight function is known to be generally a more powerful statistic test due to its emphasis on the tail disparities between the EDF and the expected cumulative distribution function. In order to facilitate the computation of such a test, it is utilized the following equation:

$$A_n^2 = -n - \frac{1}{n} \sum_{i=1}^{n} (2i - 1)[\log J(x_i) + log\{1 - J(x_{n+1-i})\}] \tag{21}$$

3 The Industrial Unit and the Process

The case study is focused on one automotive supplier enterprise, essentially, one industrial location – a Portuguese automotive SME. The organization runs in two shifts of 8 h each per day. It provisions the key Original Equipment Manufacturer (OEM) of the sector with welded assemblies, stamped metallic pieces, subassemblies, chassis and lately, with some precision pieces. The factory is specialized in diverse processes of manufacture, such as automatic and manual welding of structures or minor components, surface treatment and a diverse variety of stamping processes. The production process of every product trails a specific and individual flow, as every stamped piece needs a particular tool. The studied stamped piece coming out of the hydraulic press can be seen in Fig. 1.

In case of such types of processes and with the purpose of having a superior quality control at several levels the management applied a broad variety of SPC tools. Throughout the production process of the metal products the SPC is demanded by the client as an obligatory instrument. The gathering frequency of data for measurement and inspection is set on a sample of 5 pieces per shift. The required measurements are performed with the aid of a coordinate measuring machine (CMM). For measuring the stamped piece, a control chart ($\overline{X}$, R) is utilized. As a goal, the dimensional feature of 89.8 mm with an inferior and superior tolerance of 0.1 mm is set. The results of the SPC tool after 50 compiling and measuring the 50 collected samples can be witnessed in Fig. 2. As indicated by this figure, it is possible to deduce that the process does not show any special causes or unexpected behaviours and thus is deemed under control.

A graphical technique intended to identify fundamental eccentricities from normality is the cumulative frequency distribution with regression analysis. This type of test related to the SPC control chart at the industrial facility demonstrates that the analyzed data tails a normal distribution as can be seen in Fig. 3. It can be observed that the detected cumulative frequency fallows the expected value of normal distribution. The regression analysis shows that R^2 is close to 1 since the higher the R^2, the better the model fits the studied data.

4 Result Analysis

For the studied case the C_{pk} and other process capability ratios can be observed in Table 1. The minimum C_{pk} recommended for two-sided specifications is 1.33 [15] and the results can be seen in Table 1 indicates that this requirement is properly followed. Besides, all the results from Figs. 2 and 3 indicate that the process is capable. Yet, a confirmation has to be made through K-S, Shapiro-Wilk and Anderson-Darling tests whether the recorded data follows a normal distribution.

It was defined in this study that null hypothesis is that the data are normally distributed while the alternative hypothesis is that the data are non-normal. The significance level for Anderson-Darling and Shapiro-Wilk was set to 0.05 meaning that if the result of the p-value is inferior to 0.05 then the data do not follow the normal distribution. After all of the three proposed goodness of fit tests were performed on the same data of the control chart several results were obtained.

Fig. 1. The studied stamping press.

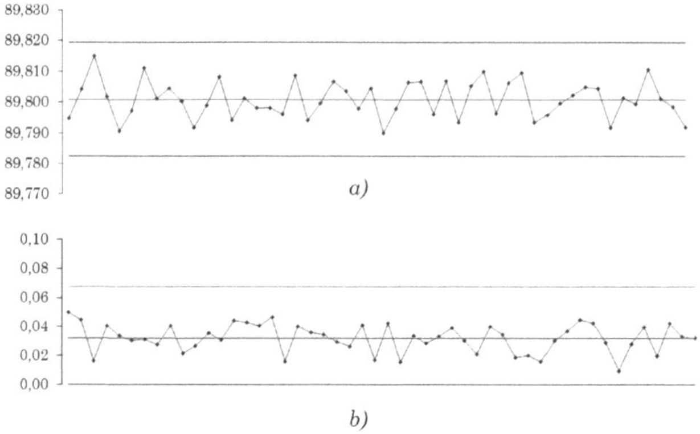

Fig. 2. Variables control chart of 50 samples of (a) mean and (b) variance.

The K-S test shows that the observed cumulative frequency narrowly fallows the expected value of the normal distribution as can be seen in Fig. 4. The results of the Shapiro-Wilk test also indicate that the data is well-modelled by a normal distribution since the p-value is 0.759 which is > 0.05, as can be observed in Fig. 5.

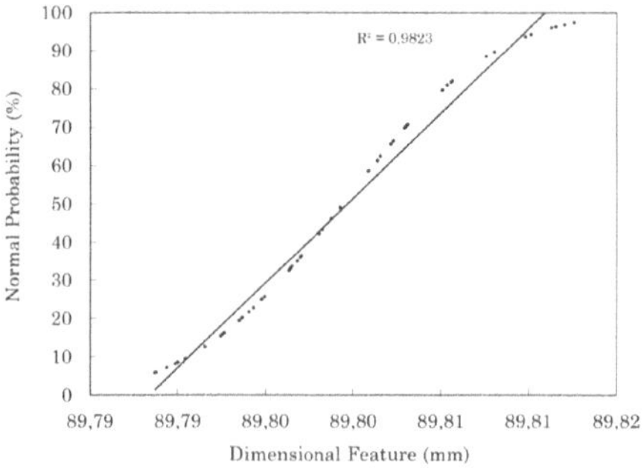

Fig. 3. Cumulative frequency distribution.

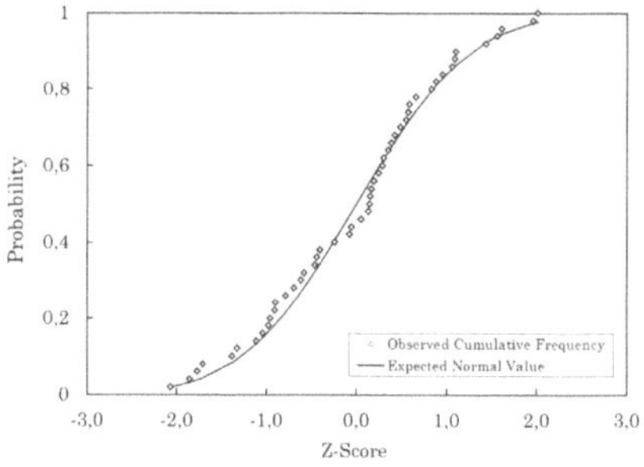

Fig. 4. The results of Kolmogorov-Smirnov test.

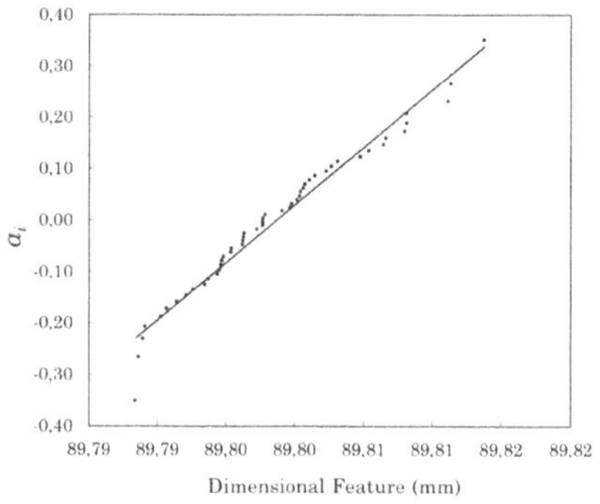

Fig. 5. The results of the Shapiro-Wilk test.

In this figure it is possible to observe that the a_i of the dimensional feature follows a straight line without having records with skewed deviations. Finally, as depicted in Fig. 6, the results obtained from the Anderson-Darling show that the data follow a normal distribution as well since the Z-Value does not show significant deviations. A comparison was made after all of the goodness of fit tests were performed on the same registered values. By observing all Figs. 4, 5 and 6 it is assessed that the observed data does not show significant deviations from the mean which usually indicates non-normality.

The comparison of the obtained results of each test can be observed in Table 2. In the K-S test the calculated $D_{n,\alpha}$ is 0.089 and by comparing it with the critical value from the from Kolmogorov distribution table [17] which is 0.22585 it can be deduced that the data follow a normal distribution. Furthermore, since the p-values from the Anderson-Darling and Shapiro-Wilk are significantly larger than 0.05, the null hypothesis is accepted that the data comes from a normal distribution. In the case of the Anderson-Darling test the p-value is 0.348 which is meaningfully larger than 0.05 while in the case of the Shapiro-Wilk test the p-value is even larger - 0.759.

All four normality tests confirm that the collected data come from a normal distribution. Therefore, such an outcome signifies that the process is in control. Such results give the quality engineer an increased reassurance in the control chart at his

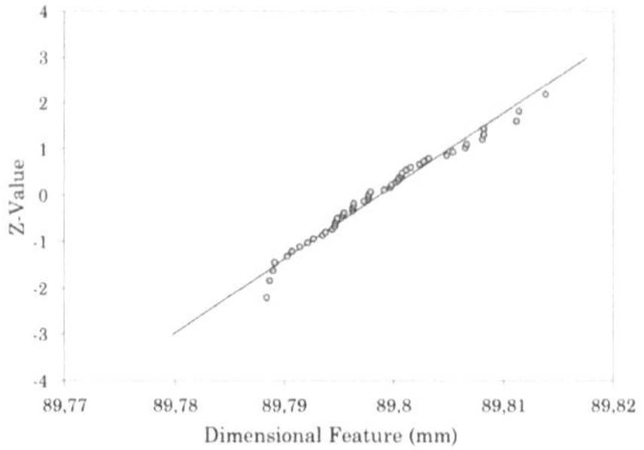

Fig. 6. The results of the Anderson-Darling test.

Table 1. C_{pk} and other process capability ratios.

Number of readings	250
Lower spec limit (*LSL*)	89.700
Nominal	89.800
Upper spec limit (*USL*)	89.900
Out of control limits	0
Average range (*R*)	0.05
Upper capability index (C_{pu})	1.63
Lower capability index (C_{pl})	1.66
Capability index (C_p)	1.64
Process capability index (C_{pk})	1.63

disposal since it allows to affirm with a high accuracy that the analyzed data comes from a normal distribution. Being aware of which statistical tool to employ in a given situation will guarantee an accurate monitoring of process stability. It will also increase the chance of eliminating invalid results and wasted effort, paving the way for proper opportunities for significant improvement. This outcome allows the industrial unit to reinforce the control charts currently at their disposal and by doing so increases the chances to detect nonconformities. This, in turn, allows the management of the factory to decrease the unnecessary waste and therefore increase the overall sustainability of the enterprise.

Table 2. Results of goodness of fit test.

	Regression analysis	Kolmogorov-Smirnov	Anderson-Darling	Shapiro-Wilk
Test	Is R^2 far from 1?	Is $D_n > 0.2259$?	Is $p < 0.05$?	Is $p < 0.05$?
R^2	0.982	–	–	–
D_n	–	0.089	–	–
p-Value	–	–	0.348	0.759
Reject?	**No**	**No**	**No**	**No**

5 Conclusions

In this paper a study of an automotive SME in Portugal in which SPC is applied was made. The normality test of the SPC utilized at the SME is the cumulative frequency distribution with regression analysis. In the analyzed case the SPC method indicated that the process is centred and satisfies the compliance requirement and the regression analysis test shows that the documented records reflect a normal distribution. Therefore, in order to estimate more accurately if the tested data came from a normally distributed population, three of the most advanced goodness of fit tests were performed on the same collected values: Kolmogorov-Smirnov, Anderson-Darling and Shapiro-Wilk. Then, the results of all tests were analyzed, discussed and compared. Since these tests are more sensible to detect most departures from normality, they allow a more accurate assessment of the collected data which in turn increases the confidence in the control chart. The results showed that all four normality tests confirm that the collected data come from a normal distribution. Such results gave the management of the industrial unit a greater array of tools to employ in many situations which in turn assures an accurate monitoring of process stability. The reinforcement of the control chart with these statistical tools increases the chance of eliminating incorrect results, increasing the occasions for significant improvement. Overall, such a solution allows improving the accuracy of the statistical process control chart and its application can be intended to all fields that require manufacturing.

Acknowledgments. Radu Godina would like to acknowledge financial support from Fundação para a Ciência e Tecnologia (UID/EMS/00667/2019).

References

1. Alaskari, O., Ahmad, M.M., Pinedo-Cuenca, R.: Development of a methodology to assist manufacturing SMEs in the selection of appropriate lean tools. Lean Six Sigma J. **7**, 62–84 (2016)
2. Mitra, A.: Fundamentals of Quality Control and Improvement. Wiley, Hoboken (2016)
3. Godina, R., Matias, J.C.O.: Quality control in the context of industry 4.0. In: Reis, J., Pinelas, S., Melão, N. (eds.) Industrial Engineering and Operations Management II, pp. 177–187. Springer, Cham (2019)
4. Chiarini, A.: Japanese total quality control, TQM, deming's system of profound knowledge, BPR, lean and six sigma: comparison and discussion. Lean Six Sigma J. **2**, 332–355 (2011)
5. Sinclair, K., Sadler, B.: Going beyond SPC—why we need statistical thinking in operations such as carbon plants. In: Tomsett, A., Johnson, J. (eds.) Essential Readings in Light Metals, pp. 365–370. Springer, Cham (2016)
6. Schmidt, C., Li, W., Thiede, S., Kara, S., Herrmann, C.: A methodology for customized prediction of energy consumption in manufacturing industries. Int. J. Precis. Eng. Manuf.-Green Tech. **2**, 163–172 (2015)
7. Azizi, A.: Evaluation improvement of production productivity performance using statistical process control, overall equipment efficiency, and autonomous maintenance. Procedia Manuf. **2**, 186–190 (2015)

8. Cheng, Z.: Mean shifts diagnosis and identification in bivariate process using LS-SVM based pattern recognition model. Int. J. Ind. Eng.: Theory Appl. Pract. **20**, 453–467 (2013)
9. Montgomery, D.C.: Statistical Quality Control. Wiley, Hoboken (2012)
10. Zhao, X.: A process oriented quality control approach based on dynamic SPC and FMEA. Int. J. Ind. Eng.: Theory Appl. Pract. **18**, 444–451 (2011)
11. Godina, R., Rodrigues, E.M.G., Matias, J.C.O.: An alternative test of normality for improving SPC in a Portuguese automotive SME. In: Viles, E., Ormazábal, M., Lleó, A. (eds.) Closing the Gap Between Practice and Research in Industrial Engineering, pp. 277–285. Springer, Cham (2018)
12. Godina, R., Pimentel, C., Silva, F.J.G., Matias, J.C.O.: Improvement of the statistical process control certainty in an automotive manufacturing unit. Procedia Manuf. **17**, 729–736 (2018)
13. Godina, R., Matias, J.C.O.: Improvement of the statistical process control through an enhanced test of normality. In: 2018 7th International Conference on Industrial Technology and Management (ICITM), Oxford, UK, pp. 17–21 (2018)
14. Addeh, J., Ebrahimzadeh, A., Azarbad, M., Ranaee, V.: Statistical process control using optimized neural networks: a case study. ISA Trans. **53**, 1489–1499 (2014)
15. Godina, R., Matias, J.C.O., Azevedo, S.G.: Quality improvement with statistical process control in the automotive industry. Int. J. Ind. Eng. Manag. **7**, 1–8 (2016)
16. Zhang, J.: Powerful goodness-of-fit tests based on the likelihood ratio. J. Roy. Stat. Soc.: Ser. B (Stat. Methodol.) **64**, 281–294 (2002)
17. O'Connor, P.D.T., Kleyner, A.: Appendix 3: Kolmogorov-Smirnov Tables. In: Practical Reliability Engineering, pp. 455–456. Wiley (2011)
18. Kim, N.: The limit distribution of a modified Shapiro-Wilk statistic for normality to type II censored data. J. Korean Stat. Soc. **40**, 257–266 (2011)
19. Heo, J.-H., Shin, H., Nam, W., Om, J., Jeong, C.: Approximation of modified Anderson-Darling test statistics for extreme value distributions with unknown shape parameter. J. Hydrol. **499**, 41–49 (2013)
20. Shin, H., Jung, Y., Jeong, C., Heo, J.-H.: Assessment of modified Anderson-Darling test statistics for the generalized extreme value and generalized logistic distributions. Stoch. Environ. Res. Risk Assess. **26**, 105–114 (2012)

Energy LCA-Oriented Sustainability Analysis Assessment Approach for Visualization of Energy-Efficient Manufacturing

Aldona Kluczek and Bartlomiej Gladysz(✉)

Faculty of Production Engineering, Institute of Production Systems Organization, Warsaw University of Technology, 86 Narbutta Str., 02-524 Warsaw, Poland
bartlomiej.gladysz@pw.edu.pl

Abstract. In recent years, manufacturing processes have interacted with sustainability which are implemented in the cost-effective ways to minimalize energy, decrease negative impact on the environment and are safe for society. However, the attention has been on separate sustainability assessment methods combining energy Life Cycle Analysis (LCA) to evaluate environmental impact, Life Cycle Cost to analyze costs, and Social Life Cycle Assessment to consider energy in human resources through mapping them in "energy LCA-based value stream map" (eVSM). In this paper, the energy LCA-oriented sustainability assessment approach is used to visualize the energy sustainability of the milk processing in terms of technologies. It enables interventions in the considered technologies through the effective analysis of energy flow. The results pictured that the appropriate integration of technologies into eVSM offers LCA-indicators database regarding the energy sustainability.

Keywords: Energy efficient manufacturing · LCA-based energy value stream map · Life cycle cost · Social life cycle analysis · Sustainability assessment

1 Introduction

Progress towards improving energy efficiency performance in manufacturing requires the application of new approaches for an assessment considering energy as an independent dimension within the TBL (triple bottom line) or sustainability. Sustainability addressing economic growth, environmental protection, and social equity requires assessment tools and approaches, often integrated assessment methodologies which are different in their interpretation [1]. A choice of assessment methods in which sustainability assessment approach will be used depends on decision-makers [2]. A sustainability challenges can exist in energy initiatives emphasizing the integration of energy into sustainability assessment methods, adapting new more efficient manufacturing technologies, which are affected by rising energy prices [3]. Hence, energy-efficient manufacturing can have implications for decision-making processes and a broader sustainability interpretation [2]. Energy audits are as the most energy-efficiency measurement method but not sufficient data to provide impact evaluation. In addition, even when necessary data are available, they are limited to the highest levels of

M. Majewski and W. Kacalak (Eds.): IIRTS 2019, LNME, pp. 61–72, 2020.
https://doi.org/10.1007/978-3-030-37566-9_6

aggregation [4, 5] it is difficult to calculate energy-related indicators which should be consistent and applicable for energy flow analysis.

Progress toward energy efficiency or energy sustainability requires integrated assessment methodologies [1]. Available sustainability assessment methods, like life cycle analysis (LCA) are incomplete and do not comprehensively assess some sustainability aspects. LCA, LCC (life cycle costing), SLCA (social LCA) are selected for the assessment hardwood lumber production/technologies focusing on the interaction between methods elements. Energy stream map should give transparent results, but they do not provide impact evaluation combining LCA, LCC, and SLCA. Hence, there is need to extend standardized approach used in sustainability manufacturing assessment that give more both transparent, consistent and comparable indicators that are applicable for energy flow analysis. From the methodological point of view, energy consumption in the industry sector, besides level of information is available, requires taking into consideration issues that can be addressed by developing indicators at different levels of aggregation, consistent boundaries and allocation concerns related to arising energy consumption and carbon dioxide emissions indicators [6].

The goal of the paper is to analyze energy flow by proposing "energy LCA-based value stream map" that can be applied to efficient assessment of energy sustainability and get insights into potential improvements of industrial energy technologies efficiency. Due to simplicity and user-friendly method, energy-related value stream map (eVSM) is used in this paper. This tool merges environmental, economic and social indicators providing an energy LCA-based sustainability assessment framework.

2 Literature Review

Up to now, research in sustainability area has been the hot topic of the academia. In manufacturing, energy efficiency can be improved by three different approaches as follows [6]:

- Energy savings by management;
- Energy savings by technologies;
- Energy saving by policies/regulations.

The energy-led value stream map (eVSM) visualizes environmental, economic, and social "energy sustainability performance" in terms of three assessment methods: LCA, LCC, SLCA. Also, energy efficiency has environmental and economic roots, but it is also associated with social energy-related factors. While energy relates to CO_2, therefore any investments to reduce energy consumption and carbon dioxide simultaneously could be treated as energy-environment efficiency for the analyzed process.

Current available methods, based on LCA are treated as tools for sustainability assessment [1], however they are not without their limitations. Therefore, combined LCA methods provide most comprehensive energy assessment for sustainability. Some researchers provided a high range of energy indicators, but avoiding consideration of economic and social impacts. Here, Verma and Sharma [7] used the concept of energy value stream mapping for non-value-added energy consuming processes, addressing the overall productivity. In industrial processes, Helleno et al. [8] proposed the model incorporating environmental and social indicators into value stream mapping approach (VSM). Similar, the VSM developed by Muller et al. [9] provides additionally value-adding streams in production and logistics in respect of time- and energy-consumption. Once, many energy-efficiency approaches [10] and sustainability assessment measures are presented [11], energy LCA-oriented sustainability stream roadmap of energy efficiency is not realized in industrial plants. A comprehensive energy survey of an industrial-scale long-life milk processing plant was carried out only [12]. LCA method is limited to environmental assessment only, hence requires "a novel approach that complements environmental and economic considerations" [13]. It is addressed in this research with a graphical visualization of energy flow on LCA-based value stream map. The presented map depicts combination of environmental, economic and an additional dimension of social issue. In recent years the social dimension has gained greater attention and has been increasingly included into manufacturing processes [14, 15] focusing on energy flow analysis.

Here, comparative energy life cycle assessment (eLCA) is an approach to assess the direct and indirect environmental effects of industrial operations on human health, and through economic benefits simultaneously.

While environmental sustainability assesses carbon footprint by minimalizing greenhouse gasses emissions (social sustainability), and their economic impact (stimulating economic effects of technologies), energy LCA would assess impact on energy use. However, it is still missing a successful integration of human factor insights into energy efficiency measures.

With regard to the milk processing, technologies mainly address separate technology improvement measures for energy savings, but there is no methodology for assessing and picturing the integrated effects from a combined analysis of energy, pollutions, and human factor so far. Taking eVSM, it integrates the environmental and economic and social sustainability dimensions to enable energy LCA-oriented sustainability of energy-efficiency of technologies in dairy industrial plants.

3 Methods

LCA-objective energy (value) stream mapping using the basic Value Stream Methodology is developed in this paper to present energy flow, emissions, cost and social energy contributions. Which technologies of dairy operation are essential from an environmental and economic, social point of view? Answering the question is possible to find in a clear way on the eVSM map using LCA-based indicators which are put in relation to energy-related emissions and cost. Some assumption should be distinguished. The relevance of a method depends on the availability of data. Time is ignored through the eVSM. Different levels of aggregation of energy related indicators are included. Consistent boundaries and allocation concerns are related to arising energy consumption (based on [6]).

Depending on the level of information available, the energy sustainability assessment framework comprises scenario analysis through energy-related indicators within LCA-based methods: energy life cycle assessment, life cycle costing, social life cycle analysis, as which are used to assess the most sustainable energy efficiency technologies in manufacturing (see Fig. 1).

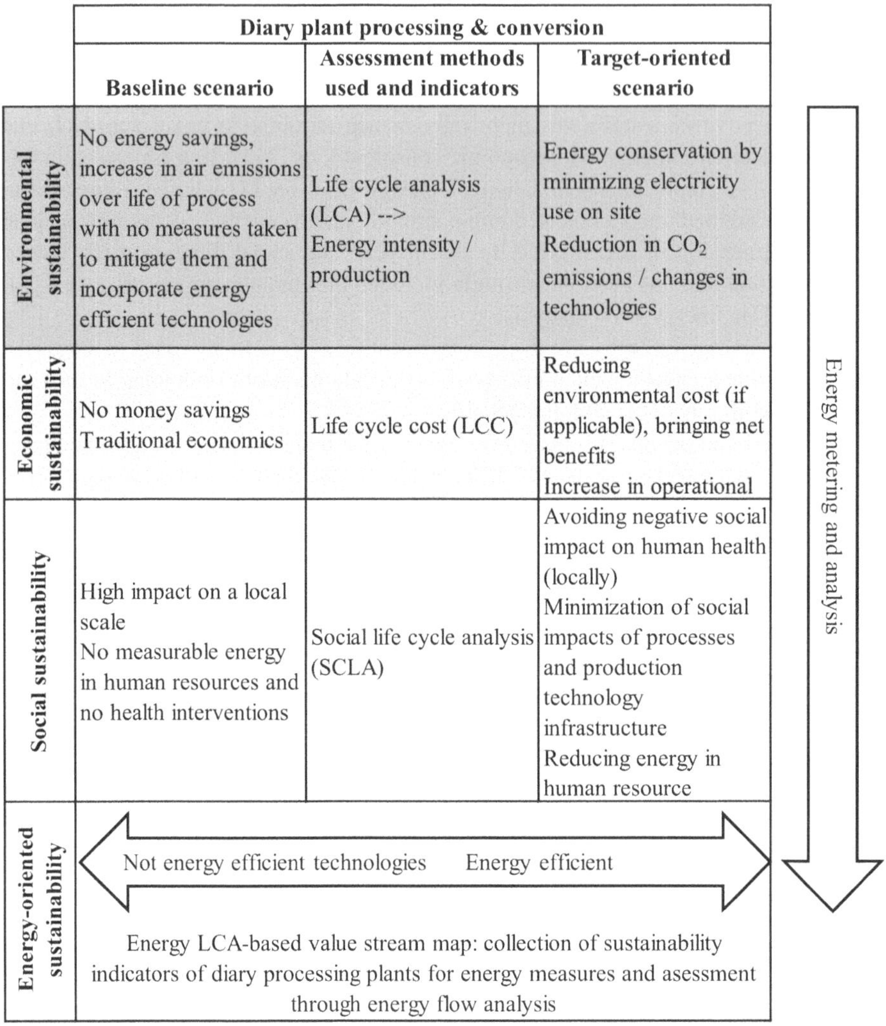

Fig. 1. Schematic sketch of energy LCA-oriented sustainability analysis.

This method is not standardized, because it refers to allocation of the eVSM inputs (energy consumption) to current, single, and not energy efficient technologies compared to allocation to energy inputs of recommended one (energy efficient). Energy LCA-oriented indicators are depicted here for visualizing energy sustainability production, described in Sect. 3.2. Very often, indicators by the optimal solution at the first level (baseline scenario) yield the reference point to the second level, leading them to sustainable solutions (improvement scenario). Hence, possible two scenarios using eVSM will be developed.

3.1 Process Background

Raw and processed milk is blended with other ingredients (sugar, flavours, etc.) to get the final product. Since the product blending is finished it is thermally processed to provide appropriate quality products (as fluid milk, yogurt or other dairy products) and packed into final form (bags, containers, boxes, etc.). The generic process is designed, because it covers the wide range of product manufacturing activities (Fig. 2).

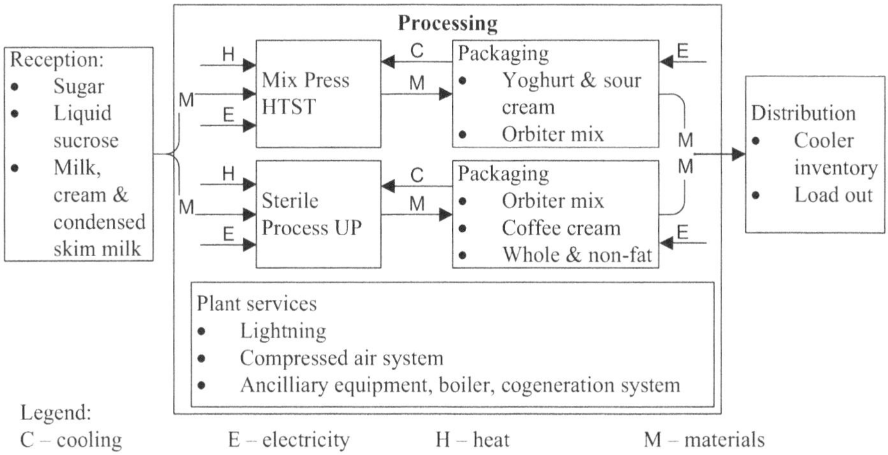

Fig. 2. General schematic view of the milk processing.

Major energy consuming equipment refers to air compressors, ammonia compressors, pumps, lighting, motors in food processing lines, and natural gas, i.e. steam for process heating, facility heating.

Dairy industry has a great potential for improving the efficiency of technologies and the performance and sustainability parameters of dairy processing plants. High share of diary operations for climate change and energy is provided by natural gas boiler (heat). The diary plant manufactures 27,914,796 kg of milk's products per year.

3.2 Building-LCA Indicators for Energy-Efficient Manufacturing

Selected non-traditional indicators (input-outputs), such as used in LCA analysis for each considered technology expressed as per dm^3 of total production are allocated must be considered when assessing energy-led sustainability of energy-efficient manufacturing in terms of technologies (Fig. 1). There are energy LCA based indicators represented by sustainable assessment methods in terms of sustainability pillars as follows:

- Energy intensity/production [kWh/dm^3 of fluid milk],
- Global Warming Potential in terms of release of CO_2 due primarily to the human use of fossil fuels, which releases CO_2 and other greenhouse gases into the air [16],
- Energy per employee-year consumed per total annual production [kWh/dm^3],
- Life cycle cost [$/dm^3] involving capital investment (I) and operational & maintenance cost (O&M), calculated as follow: (I + O&M) – LCR, where LCR – revenues from energy savings. The O&M calculation is not included in this paper.

4 Assessment Through Mapping a Building-LCA Indicators

To assess a feasible baseline and improvements actions for increasing efficiency of technologies through milk processing, a target-oriented scenario is developed. This shift can be accelerated with recommendations that look at both the current and future and target oriented state emphasizing technical improvement such as: (1) reduction of the energy intensity by upgrading and deployment of the best available technologies, (2) reduction emissions. The results of the comparisons of data through the energy stream maps using the comparative LCA-based indicators are outlined in Figs. 3 and 4. These actions improve process performance and plant efficiency. It poses a challenge on ensuring compliance to standards of diary operations on the eVSM. Distribution of the total energy consumption and its LCC, presented in both scenarios is outlined in Fig. 5.

5 Results

The study has shown that technologies have strong energy impact on the raw milk production. Considering baseline scenario, cooling and refrigeration has the greatest contribution to the energy assessment (32% of energy use) followed by material processing (30.6%), pumps respectively (13%). The target-oriented scenario generates energy efficiency improvements. By upgrading considered technologies (baseline) into energy efficient ones (Fig. 5), there is a potential saving of energy and hundreds of thousands of dollars per year, e.g. lightings will save US$ 53,660 (4.7% of total energy use) reducing energy by 4% in human resources required for conducting non-value-added activity (controlling milk processing and the material movement). According to the outcomes of the baseline study, the largest energy consumption appeared in cooling and refrigeration process, accounting for 32.2% of total energy use.

The modifications in this process through heat recovering from ammonia cooling towers improved the energy sustainability level by 2%, reducing CO_2 emission about 21%.

The improvement scenario in Fig. 6 shows the target CO_2 emission (+) comparing this with the potential percentage differences as a result of the technological intervention between the scenarios. When considering as a whole, proposed technologies can reduce carbon dioxide emissions by 27% of current production per year. The biggest CO_2

emissions occurred in the steam generation. By installation of cogeneration system, Global Warming Potential (GWP) was decreased by 54% of CO_2 emitted (Fig. 6).

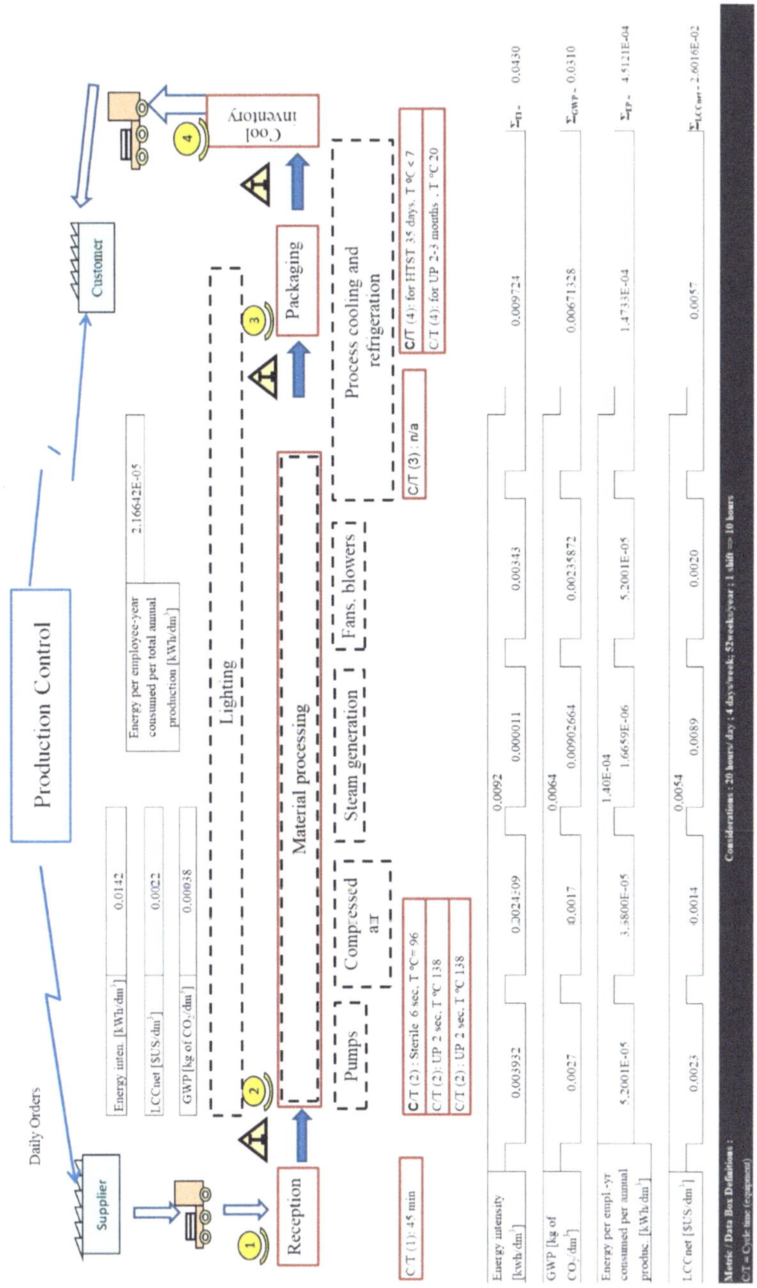

Fig. 3. eVSM in baseline scenario.

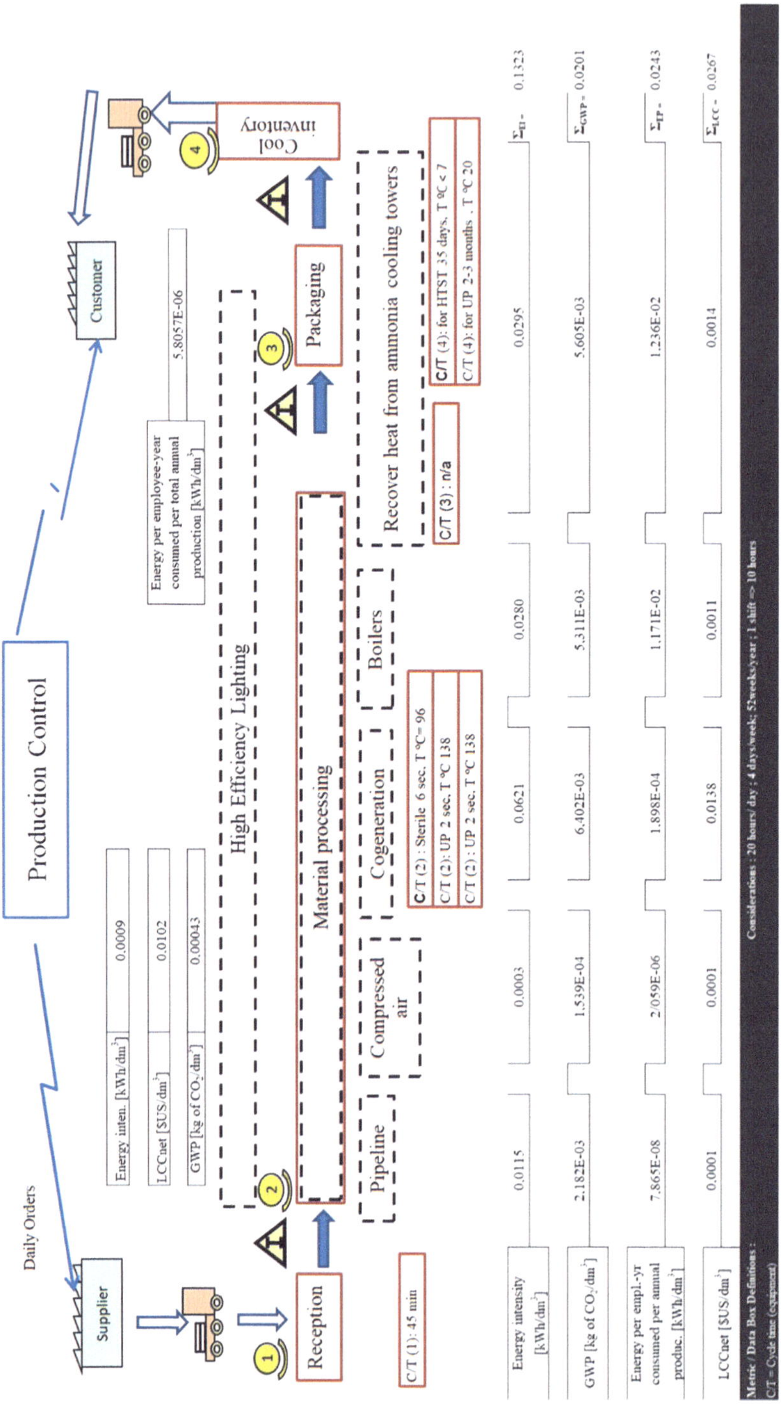

Fig. 4. eVSM in improvement scenario.

Technology	Energy usage [kWh/yr]	% of total use	CO_2 eq. [kg/yr]	LCCnet per year	Proposed technology	Energy savings per yr	Reduction of CO_2 emission [kg/yr]	Energy savings [$/yr]	Implementation cost [$]	LCCnet per yr
Compressed air	340000	8	234507	38420	Cogeneration system	1750000 kWh -7070 MMBtu	834610	166000	5000000	374440
Fans, blowers	480000	11	331122	54240	High efficiency lightning	53360 kWh	36287	6060	9750	276361
Material processing	1290000	31	889041	145770	Duct outside air to compressor intakes	19020 kWh	13608	2130	25000	1894
Pumps	550000	13	381018	62150	Recover heat from ammonia cooling towers	6195 MMBtu	326587	28000	50000	37811
Cooling and refrigeration	1360000	32	938936	153689	Preheat boiler inlet air with exhaust air	5870 MMBtu	312979	26800	40000	30025
Lightning	200000	5	137892	22600	Pipeline insulation	2411 MMBtu	13608	11000	4000	3018
Total electricity		100	2912517	476869	**Total electricity**	1822680 kWh				
Steam generation	52500	100	2785057	239925	**Total natural gas**	21546 MMBtu				
Total	6421000 52500		5697574	716785	**Total**	1882690 kWh 14476 MMBtu	1537678	60824030	606250	723551

Energy consumption at the diary plant under study

	Usage	Cost [$]
Electricity	4225420 kWh	479242
Natural gas	52500 MMBtu	234814
Total	66918 MMBtu	719060

Fig. 5. Distribution of the total energy consumption and its efficiency.

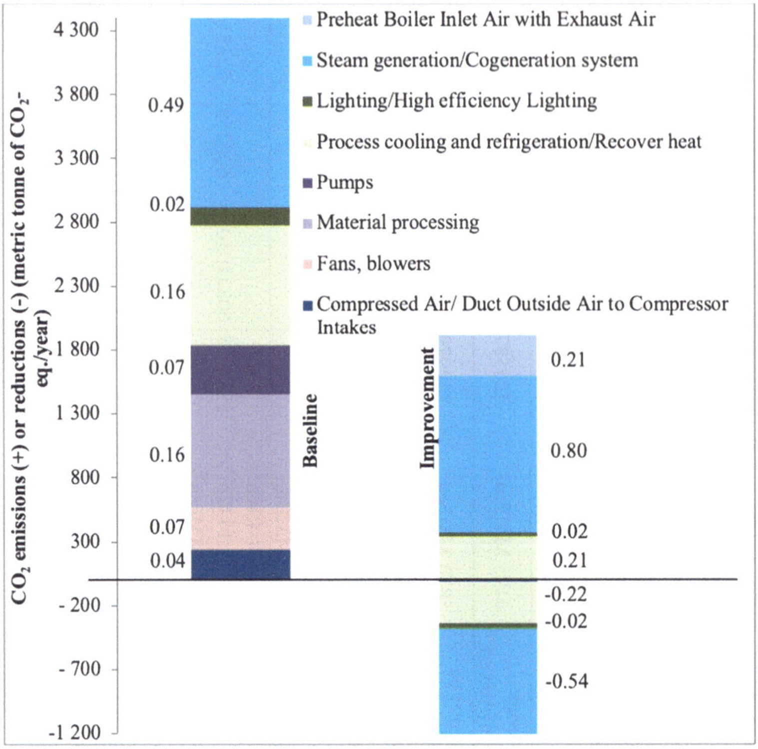

Fig. 6. Comparison of CO_2 emissions for both scenarios.

LCCs were impacted on manufacturing performance by energy technologies. It is expected the greatest economic improvements with the cogeneration system, decreasing costs by 33% against baseline scenario followed by installing high efficiency lighting, therefore reducing LCC by 25%.

6 Conclusions and Discussion

In this article, the connotation energy efficiency and sustainability of manufacturing technologies could be visualized. Then, assessment of milk processing in terms of used technologies based on energy flow, emissions and cost were introduced in more detail. The contribution is placed in the broader context of the modern paradigm of sustainability, namely energy sustainability and their implication for assessment industrial plant. Different interpretations based on energy flow analysis can have implications for decision-making processes [2]. As stated by Verma and Sharma [7], the eVSM can be applied not only for diagnostic purposes but also for energy costing and saving measures with regards to social issue.

The use of LCA-oriented sustainability for energy-efficient manufacturing could be applied for the following purposes: (1) identification of energy environmental, economic and social issues; (2) comparing alternative performance of technologies, simultaneously building the same system boundaries, identifying energy-efficient manufacturing processes. Therefore, the system boundaries will include a process phase (operational energy use and technologies replacement), not involving the product stage (raw material supply, transport and manufacture of products used in the plant). With the improvement scenario, the excessive energy and life cycle cost per volume associated with energy can be minimized. Through energy and emissions visualization, eVSM can allow managing energy-related indicators, thus moving from small improvements toward greater enhancements to energy-efficient manufacturing. Additionally, improvements that are integral to the environmental management standard enable plant's staff also to minimalize environmental impact (save energy), reduce energy cost and environmental cost fees. Personnel who is engaged in energy saving practices and working toward improvement of energy efficient technologies, they are more involved overall. The proposed indicators might be useful in plant's strategic planning, including decision- and policy-making with respect to the development of appropriate sustainable production management systems. Further studies will intend to broaden this research using energy performance analysis of considered technologies as a standard that can be used to best evaluation. Further research will focus on examination of proposed approach on other relevant industrial examples, to constitute a basis for the development of standardized approach to energy efficiency mapping.

References

1. Sala, S., Farioli, F., Zamagni, A.: Life cycle sustainability assessment in the context of sustainability science progress (part 2). Int. J. Life Cycle Assess. **18**(9), 1686–1697 (2012)
2. Ness, B., Urbel-Piirsalu, E., Anderberg, S., Olsson, L.: Categorising tools for sustainability assessment. Ecol. Econ. **60**(3), 498–508 (2007)
3. Schrettle, S., Hinz, A., Secherrer-Rathje, M., Friedli, T.: Turning sustainability into action: explaining firms' sustainability efforts and their impact on firm performance. Int. J. Prod. Econ. **147**(Part A), 73–84 (2014)
4. Phylipsen, G.J.M.: Energy efficiency indicators best practice and potential use in developing country policy making. Phylipsen Climate Change Consulting (2010). Accessed 01 Apr 2019
5. OECD/IEA: Energy Efficiency Indicators: Essentials for Policy Making. OECD/IEA (2014). Accessed 01 Apr 2019
6. Abdelaziz, E.A., Seidur, R., Mekhilef, S.: A review on energy saving strategies in industrial sector. Renew. Sustain. Energy Rev. **15**, 150–168 (2011)
7. Verma, N., Sharma, V.: Energy value stream mapping a tool to develop green manufacturing. Procedia Eng. **149**, 526–534 (2016)
8. Helleno, A.L., Moraes, A.J.I., Simon, A.T.: Integrating sustainability indicators and Lean Manufacturing to assess manufacturing processes: Application case studies in Brazilian industry. J. Clean. Prod. **153**, 405–416 (2017)
9. Muller, E., Stock, T., Schilling, R.: A method to generate energy value-streams in production and logistics in respect of time- and energy-consumption. Prod. Eng. Res. Dev. **8**(1–2), 243–251 (2014)

10. Thollander, P., Palm, J.: Improving Energy Efficiency in Industrial Energy Systems. An Interdisciplinary Perspective on Barriers, Energy Audits, Energy Management, Policies, and Programs, pp. 15–34. Springer, London (2013)
11. Moldavska, A., Welo, T.: On the applicability of sustainability assessment tools in manufacturing. Procedia CIRP **29**, 621–626 (2015)
12. Soufiyan, M.M., Aghbashlo, M., Mobli, H.: Exergetic performance assessment of a long-life milk processing plant: a comprehensive survey. J. Clean. Prod. **140**(Part 2), 690–697 (2017)
13. De Benedetto, L., Klemes, J.: The environmental performance strategy map: an integrated LCA approach to support the strategic decision-making process. J. Clean. Prod. **17**, 900–906 (2009)
14. Cao, Y., Wang, S., Zhou, L.Y.J.: A social sustainability assessment model for manufacturing systems based on ergonomics and fuzzy inference system. In: Setchi, R., Howlett, R., Liu, Y., Theobald, P. (eds.) Sustainable Design and Manufacturing, pp. 639–648. Springer, Cham (2016)
15. Sutherland, W.J., Richer, J.S., Hutchins, M.J., Dornfeld, D., Dzombak, R., Mangold, J., Robinson, S., Hauschild, M.Z., Bonou, A., Schönsleben, P., Friemann, F.: The role of manufacturing in affecting the social dimension of sustainability. CIRP Ann. **65**(2), 689–712 (2016)
16. NEF: Simple Carbon Calculator. www.carbon-calculator.org.uk. Accessed 01 Apr 2019

Assessment of the Integrity of the Object Based on the Correlation of Super-Signals

Pavel Balagurin(✉), Pavel Makarikhin(✉), and Anastasia Grigorieva(✉)

St. Petersburg State University, St. Petersburg, Russia
pavelbalag@mail.ru, makarikhin.p.a@gmail.com, a.v.grigorieva@spbu.ru

Abstract. The assessment of the integrity of the structure of the concrete mass under the influence of temperature loads was carried out. The purpose of this work is to analyse the change in the amplitude of signals propagating from two entry points into the structure and recorded by AE sensors located in different areas of the structure. The signals are clustered using the Pearson correlation coefficient. An algorithm for the formation of super-signals for each cluster is proposed. Clusters and super-signals are presented at a value exceeding the 0.9 threshold. Based on the super-signals, the state of the concrete mass in time is estimated and the compliance of the actual flow of the experiment is checked. It was possible to establish that the proposed method of analysis leads to a correct assessment of the state of the object.

Keywords: Acoustic emission · Concrete · Cracks in the object · Super-signal · Correlation

1 Introduction

One of the most promising methods of technical diagnostics is the method of acoustic emission (AE). The method is based on the registration of ultrasonic waves emitted by structural defects in their development. Along with the significant advantages of the method, such as high sensitivity, detection of developing defects, remote control, there are certain drawbacks: significant criticality to the presence of background high-frequency noise, complexity of interpretation of control results. To date, the problem of subjectivism in the procedures for interpreting the data obtained is a major constraint in the further development of the AE method.

Since the most substantial task of AE inspection is to record AE events associated with the formation and growth of cracks in concrete (place and time of formation, size estimation, evolution), it is necessary to identify these events in the general data set and to identify signals from cracks in the background of noise.

The purpose of this work is to analyse the change in the amplitude of signals propagating from the two points of entry into the structure (Waveguide No. 11 and Waveguide No. 13) and the recorded by AE sensors located in different areas of the structure, to obtain information to assess the overall structural rearrangement of the array under the influence of loads applied to it during the experiment.

M. Majewski and W. Kacalak (Eds.): IIRTS 2019, LNME, pp. 73–80, 2020.
https://doi.org/10.1007/978-3-030-37566-9_7

2 Related Works

Thus, the task of clustering is relevant for the area of AE, as it allows to obtain data sets with similar nature. There are different methods for solving it and different areas of application of the results. So, in [1] for search of similar signals the waveform is considered. In [2] the k-average method is used for the frequency allocation function. In [3] clustering is done on the basis of a priori obtained critical intervals for values of signal characteristics. In [4] the energy of the AE signal is evaluated by means of spectral analysis. In [5], the recognition of AE sources takes place in the form of a power spectral density curve. In [6] the sliding window method is used. In [7] clustering is used to monitor the progress of material wear. In [8], clustering improves the accuracy of AE source location. The most interesting results were obtained in [9], where a set of several characteristics is considered, the interrelations between them are determined and two methods are compared: the main components method and Kohonen's self-organizing map. The method of main components is also used in [10]. In [1, 2, 9], machine learning is used. All the above mentioned methods have their advantages and disadvantages. And the fact that for so many years of research there has been no common method of grouping signals into classes, indicates that the task remains open.

3 Proposed Method for Clustering

In this paper the method of clustering of AE-signals by means of Pearson correlation coefficient is offered.

The proposed approach is tested on the data obtained from the mechanical calibration. Calibration serves for checking the functionality of the equipment before loading the object with the help of a simulator. The simulator of AE signals in this study is a click on the waveguide by a solid-state meter. This is similar to the Su-Nielsen source often used in the AE at [11, 12]. In our case each of the two calibration series consisted of 10–11 signals. Roughly speaking, the signal is an ultrasonic wave that appears inside the examined object at the moment of occurrence of a physical event (appearance of a micro-crack, growth of an existing crack, consolidation of several defects into one, mechanical click, extraneous noise, etc.). This wave propagates from the place of origin in all directions. As it progresses, it is detected by high-sensitivity sensors in the form of a set of characteristics. As a result, data from individual sensors are registered and recorded, which together provide information about the event. Sensors usually capture wave characteristics such as amplitude, duration, frequency, energy, rise time, etc.

It can be noticed in the charts and correlation matrices in the paper the numbers of signals correspond to the order of their registration, i.e. 0–10 - signals from the first series, 11–20 - from the second series.

Different types of damage and cracks generate different types of AE signals. This property of AE is used in the construction of criteria to assess the stages of destruction of an object.

Clustering of AE pulses is based on the correlation function. In mathematical statistics, correlation coefficients are traditionally used to determine the correlation between the two values. We have selected the Pearson ratio. Pearson's correlation coefficient is a method of parametric statistics that allows to determine the presence or absence of a linear relationship between two quantitative indicators, as well as to evaluate its tightness and statistical significance [13]. In this case, it reflects the similarity of the dynamics of the behaviour of the fixed characteristic of the two signals. Pearson's correlation coefficient is calculated using the following formula:

$$r(x, y) = \frac{\sum_{i=1}^{n}(X_i - \bar{X})(Y_i - \bar{Y})}{\sqrt{\sum_{i=1}^{n}(X_i - \bar{X})^2(Y_i - \bar{Y})^2}}$$

where

$$\bar{X} = \frac{1}{n}\sum_{i=1}^{n} X_i, \quad \bar{Y} = \frac{1}{n}\sum_{i=1}^{n} Y_i$$

- average values of samples, n - sample volume.

4 Proposed Algorithm for the Formation of Clusters

Previously, for each calibration experiment, the final clusters were selected manually. An algorithm is proposed to automate the selection of the most suitable clusters:

- Pearson coefficient is calculated for each signal pair. As a result, we get the matrix of Pearson coefficient values;
- for each signal (line of our table) the number of such Pearson coefficient values that exceed the threshold value is calculated, i.e. the number of similar signals is located. Similar signals form a cluster. Let us call such a number of signals the power of the cluster;
- then those clusters whose capacity is less than 3 and more than 10 are removed from consideration. We believe that clusters with power less than 3 contain erroneous signals, and clusters with power more than 10 do not reflect the real picture of the experiment, because the calibration consisted of 10 signals;
- in each cluster, by averaging all the waves it contains, we form a so-called super-signal;
- Pearson coefficients are calculated for each pair of obtained super-signals. As a result, we also get the matrix of Pearson coefficient values. If any value for a pair of super-signals exceeds the threshold value, the super-signal with the lowest power are excluded from consideration;
- for each super-signal, the products of the Pearson coefficient values are calculated;
- the received products are sorted in ascending order. As the calibration was carried out in two different parts of the facility, it is expected that there will be two clusters as well. Therefore, two of the first five super-signals with the highest capacity are selected. The first super-signal represents a series of blows on one waveguide, and the second - on the other.

5 Experiments

Detailed dimensions and scheme of the object of study (metal-concrete array) are described in the work [14].

Before applying the developed software to real experimental data, it was tested on so-called calibration data. It allowed us to know in advance where clear clusters are expected to appear, what signals should be included into them and thus check the correctness of the method's operation and absence of errors in its implementation code.

Each calibration series consisted of 10 pulses introduced into the array by two waveguides. The place of signal input via waveguide No. 11 is in the upper part of the structure, while waveguide No. 13 is in the lower part of the structure.

Many tests were performed on calibration data from different days of the experiment (total of about 20 days/experiments). The obtained results were approximately similar, that is why in the paper we limit ourselves to reviewing and analysing charts for one of the experiments which took place on November 13. These matrices and graphs of signals united into a cluster are obtained with the help of developed software that implements the proposed approach in the language of C# and R.

Tables 1 and 2 show the calculated Pearson coefficient of each super-signal with each one. Let us turn to the charts. Clusters are searched for in the multidimensional criterion space of several characteristics of AE signals. The signal sets found that belong to the same cluster is displayed in the time-amplitude section. The choice is conditioned by the fact that in this case the phase synchronization of the processes and stability of the time delay of these processes, which allow to judge whether the method has correctly defined the cluster, is most noticeable.

As it can be seen from the charts, the signals are really clearly divided into sets according to the a priori expectations. Among them, two different clusters were formed, one of which corresponds to the first source of signals (Fig. 1a), the other to the second source (Fig. 1b). These two signal divisions are obtained by setting a high threshold value of the Pearson coefficient of 0.9. As you can see, the first cluster of the program includes signals with numbers from 10 to 20, the second cluster: from 0 to 9.

This indicates a clear separation between the signals of different series. The main difference between these series was that the signals were generated in two different places of the object. Since the human factor is present in the experiment, some of the 20 signals were not as clear as others. They can be called "mistaken". They were generated at an unsuccessful imitator tilt angle or a slight impact. In the day considered, these are the signals number 1, 7.

As can be seen from the photo (Fig. 2), the cracking was active in the upper part of the object, and as a result of it there was an exit to the lead surface (metal leaks of hardened lead are visible in the photo), while in the lower part of the object there was a low intensity of cracking.

Table 1. Pearson's coefficient-based correlational matrix for super-signals from the shocks on the first waveguide

№	0	1	2	3	4	5	6	7	8	9	10
0	1	0,76	0,22	0,25	0,48	0,17	0,19	0,28	0,15	0,08	0,54
1	0,76	1	0,26	0,3	0,91	0,31	0,34	0,73	0,59	0,52	0,35
2	0,22	0,26	1	0,98	0,24	0,96	0,88	0,15	0,07	0,08	0,02
3	0,25	0,3	0,98	1	0,32	0,94	0,83	0,1	0,09	0,08	0,02
4	0,48	0,91	0,24	0,32	1	0,3	0,28	0,86	0,83	0,77	0,31
5	0,17	0,31	0,96	0,94	0,3	1	0,97	0,12	0,03	0,03	0,22
6	0,19	0,34	0,88	0,83	0,28	0,97	1	0,1	0,13	0,14	0,3
7	0,28	0,73	0,15	0,1	0,86	0,12	0,1	1	0,79	0,7	0,46
8	0,15	0,59	0,07	0,09	0,83	0,03	0,13	0,79	1	0,99	0,32
9	0,08	0,52	0,08	0,08	0,77	0,03	0,14	0,7	0,99	1	0,18
10	0,54	0,35	0,02	0,04	0,31	0,22	0,3	0,46	0,32	0,18	1

Table 2. Pearson's coefficient-based correlational matrix for super-signals from the shocks on the second waveguide

№	0	1	2	3	4	5	6	7	8	9	10
0	1	0,92	0,16	0,92	0,94	0,93	0,96	0,97	0,93	0,87	0,93
1	0,92	1	0,18	0,96	0,89	0,89	0,93	0,96	0,95	0,87	0,92
2	0,16	0,18	1	0,15	0,03	0,08	0,06	0,16	0,12	0,08	0,02
3	0,92	0,96	0,15	1	0,95	0,95	0,97	0,94	0,98	0,9	0,96
4	0,94	0,89	0,03	0,95	1	1	1	0,96	0,98	0,97	0,99
5	0,93	0,89	0,08	0,95	1	1	0,99	0,96	0,98	0,97	0,98
6	0,96	0,93	0,06	0,97	1	0,99	1	0,97	0,99	0,95	0,99
7	0,97	0,96	0,16	0,94	0,96	0,96	0,97	1	0,97	0,93	0,96
8	0,93	0,95	0,12	0,98	0,98	0,98	0,99	0,97	1	0,96	0,99
9	0,87	0,87	0,08	0,9	0,97	0,97	0,95	0,93	0,96	1	0,97
10	0,93	0,92	0,02	0,96	0,99	0,98	0,99	0,96	0,99	0,97	1

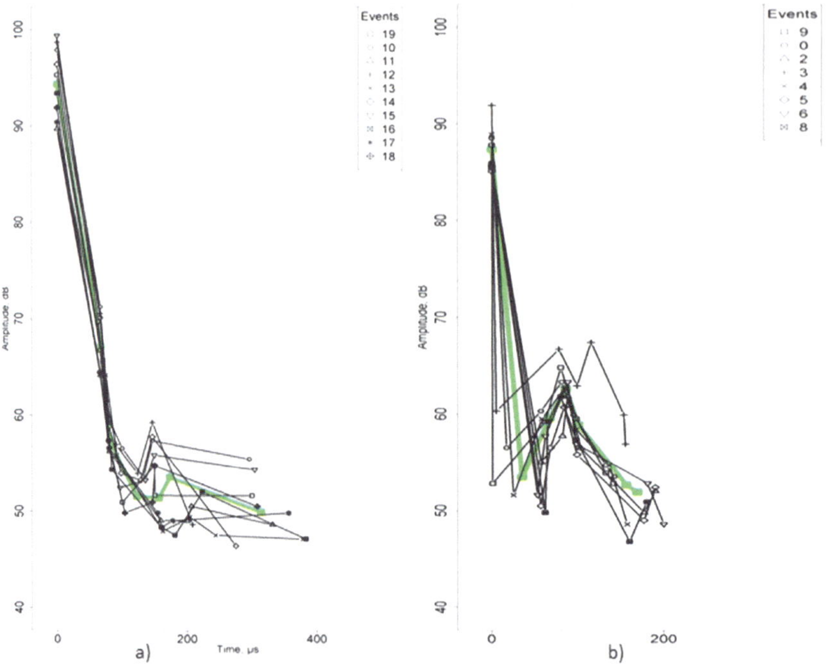

Fig. 1. Super-signal clusters corresponding to (a) the first source and (b) the second source of signals.

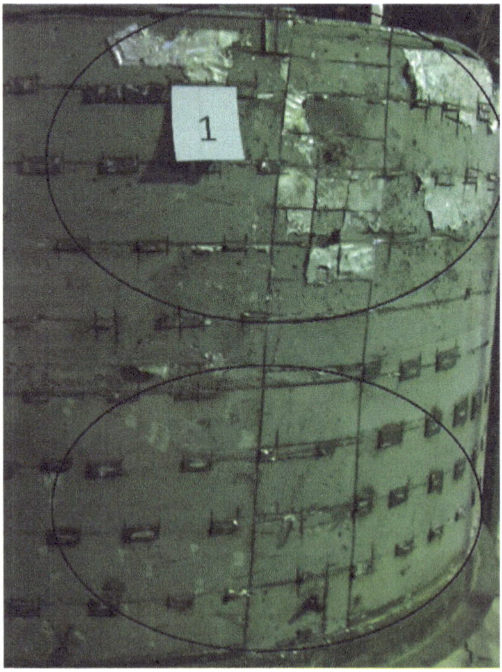

Fig. 2. Research object – metal-concrete array.

6 Conclusions and Further Research

Thus, the AE calibration method can serve as a simple and quite informative tool to control the state of the concrete mass, provided that changes are made to the procedure, eliminating some of the disadvantages of its technical implementation in the study.

Practical experiments have shown that the signals are divided into clusters according to the nature of their appearance. Software has been developed to implement the proposed approach and visualize its results in terms of amplitude-time. This work is the second step to solving an important application problem - finding an automatic way to assess the state of the object by the parameters of the flow of AE-signals. In [6] an attempt was made to solve this problem by analysing trends in the behaviour of AE-parameters graphs. Thus, the state of the object being destroyed was tracked in its entirety by means of averaging methods. Eventually, the result was recognized as positive, but not accurate enough, due to the fact that the estimates remained an element of uncertainty. Therefore, a research plan has now been drawn up to address the same task in a more subtle, composite way. This method includes: careful selection of a measure of similarity of signals (correlation coefficient), possibly for each characteristic of the signal of its measure; processing of AE data, if it is required for the possibility of applying the selected coefficients; development of a correlation function capable of steadily dividing into clusters the signals of the same nature into both test and real data, it is desirable that the clusters do not overlap with each other; study of physical properties of concrete and influence on its properties of temperature load, classification on the basis of the obtained clustering; further refinement of the correlation function by means of machine learning on the data with a known answer; fine-tuning of the model.

A type of mutual correlation function that automates the division of events into clusters:

$$F(x, y) = \sum_{i=0}^{n} g_i(x, y)$$

where x, y - waves, $g_i(x, y)$ – functions of correlation by i-th signal characteristic, n - number of functions.

In the course of further research, it is planned to carry out a similar analysis of other AE characteristics from the criterion multivariate space in order to clarify it. Choose a measure of similarity for them. Our early works allowed us to determine the first of the components, namely, to set $g_0(x, y) = r_A(x, y)$, where r_A is the Pearson coefficient applied to the amplitudes of signals x and y. In this paper we have set the second summand: $g_1(x, y) = r_{time}(x, y)$, where r_{time} is the Pearson's coefficient as applied to the time of arrival of the signal on each of the sensors that have detected the signal.

References

1. Ince, N.F., Kao, C.S., Kaveh, M., Tewfik, A., Labuz, J.F.: A machine learning approach for locating acoustic emission. EURASIP J. Adv. Signal Process. **2010**, 895486 (2010)
2. Tat, H., Wu, J., Pike, M., Schaefer, J., Pauca, V.P., Li, R., Nickerson, W.: Machine learning for acoustic emission signatures in composite laminates. In: Proceedings of the American Society for Composites - Thirty-Second Technical Conference (2017)
3. Schmidt, J., Baran, I., Ono, K.: Acoustic emission behavior of failure processes of glass-fiber laminates under complex state of loading. J. Acoust. Emiss. **23**, 173–180 (2017)
4. Kuznetsova, O.S., Sychov, S.V., Fadin, Y.A., Chulkin, S.G.: Application of acoustic emission for the study the wear of fragile materials. Sci. Educ. **3–2**(2012), 174–177 (2012). Scientific and technical statements SPbS-PU
5. Rastegayev, I.A., Merson, D.L.: Actual Problems of Strength: Monograph. EI VSU, Vitebsk (2018)
6. Belousova, V., Grigorieva, A.: Development of a software package for acoustic emission control data analysis. In: CEUR Workshop Proceedings of 3rd Conference on Software Engineering and Information Management, SEIM 2018, Russian Federation, Saint Petersburg, vol. 2135, pp. 61–66 (2018)
7. Rastegaev, I.A.: Using acoustic emission signal categorization for reconstruction of wear development timeline in tribosystems: case studies and application examples. Wear **410–411**, 83–92 (2018)
8. Barat, V.A., Alyakritskiy, A.L.: The method of statistical processing of acoustic emission monitoring data on the example of a hydrotreating reactor Mozyrskogo NPZ. Magazine "In the world of NK" **4**(42), 52–55 (2008)
9. Calabrese, L., Campanella, G., Proverbio, E.: Use of cluster analysis of acoustic emission signals in evaluating damage severity in concrete structures. J. Acoust. Emiss. **28**, 129–141 (2010)
10. Barat, V.A., Chernov, D.V., Yelizarov, S.V.: Application of data flow degradation detection methods to improve the noise immunity of the acoustic emission method. Defektoskopiya (6), 60–70 (2016)
11. Izmaylova, Ye.V., Van'kov, YU.V.: The implementation of the source of the Co-Nielsen to test the performance of acoustic emission equipment. Bulletin SUSU. Series "Computer Technology, Management, Electronics" (2014)
12. GOST R 52727-2007. Technical Diagnostics, Acoustics-Emission Diagnostics, General Requirements. Standartinform, Moscow (2007)
13. Khan, G., Shapiro, S.: Statistical Models in Engineering Problems. Mir, Moscow (1969)
14. Nefed'yev, Ye.YU., Gomera, V.P., Smirnov, A.D., Grigor'yeva, A.V., Stoyanovskiy, L.O.: Identification by the method of acoustic emission of the formation and development of cracks in the metal-concrete massif in the process of thermal testing. In: Modern engineering: Science and education: Materials of the 7th International Scientific and Practical Conference. SPb: Publishing House of Polytechnic University (2018)

Curve Curvature Analysis of a Grain Trajectories in Variable Honing of Cylindrical Holes of Thin Wall Cylinder Liners as a Honing Process Optimization Strategy

Piotr Grzegorz Sender(✉)

Department of Machine Technology and Automation of Production, Gdansk University of Technology, ul. Gabriela Narutowicza 11/12, 80-233 Gdansk, Poland
piotr.sender@wp.pl

Abstract. The main problem of honing of thin wall cylinder liners is a thermal distortion of honed holes. The higher the value of the temperature of the honed workpiece, the greater its holes deformation. The paper presents a method of reducing the temperature occurring in the honing process as a result of the application of a variable honing kinematics conditions with particular emphasis on the analysis of the effect of the value of the sum of radii of curve curvature on honed workpieces temperature. The lower value of the sum of curve curvatures radius of abrasive grain trajectories function affects the honed workpiece temperature decrease for a longer path length. The curve curvature radius of the abrasive grain paths function in the variable kinematic condition of honing of cylindrical holes can be used to improve honing of thin wall cylinder liners due to decreasing of amount temperature received in honing process.

Keywords: Honing · Variable kinematic · Abrasive grain trajectory

1 Introduction

A useful way to represent the curvature of a curve is by plotting curvature against arc length. [...] curvature can be calculated at any point on the curve from the normal and tangent vectors or from the turning angles along the curve [...] for many points on a curve would be time-consuming. It is easier to use a computationally less complex method [1]. Curve curvature *R* was calculated in the Mathcad system in many points on received previously in real honed experiment abrasive grain trajectories function of move *Z(X)*.

As shown in the literature variable kinematic condition of honing process can improve the cylindricity of honed holes [2, 4, 6]. The way of improving the surface roughness and honing efficiency was described by Buj-Corral [3]. The main problem of

M. Majewski and W. Kacalak (Eds.): IIRTS 2019, LNME, pp. 81–93, 2020.
https://doi.org/10.1007/978-3-030-37566-9_8

honing [2, 4, 6] of thin wall cylinder liners is the thermal distortion of honed holes (Fig. 1), regardless of how the holes for honing was made [5].

More thematically related and exceptionally interesting literature items are presented by author in his paper [6]. The higher the amount of the workpieces temperature T, the greater its holes deformation. The paper presents a method of reducing the temperature occurring in the honing process as a result of the application of a variable honing kinematics condition with regards to curve curvature radius *R* of abrasive grain trajectory *Z(X)*. Honed workpieces holes diameter was D = 100 mm, length of holes was l_h = 250 mm. The honing head stroke was l_s = 225 mm and material of workpieces was St 37-3 N. Research was conducted on the conventional vertical hydraulic honing machine of WMW SZS – 200. Figure 1 shows the numerical simulation of the deformation of the honed thin wall cylinder liner with different wall thickness. Different workpiece wall thickness has a different amount of cylindricity distortions [2].

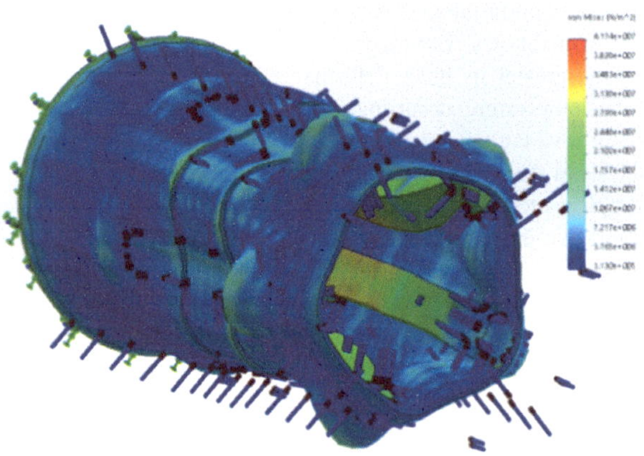

Fig. 1. Numerical simulation of deformation of honed thin wall cylinder liner with different wall thickness (temperature used in numerical simulation from real honing experiment shown on Fig. 2.

Figure 2 shows generated temperatures received in honing process (temperature of workpiece $\mathbf{T_w}$ **= 31,3 °C**, ambient temperature $\mathbf{T_a}$ **= 20 °C**).

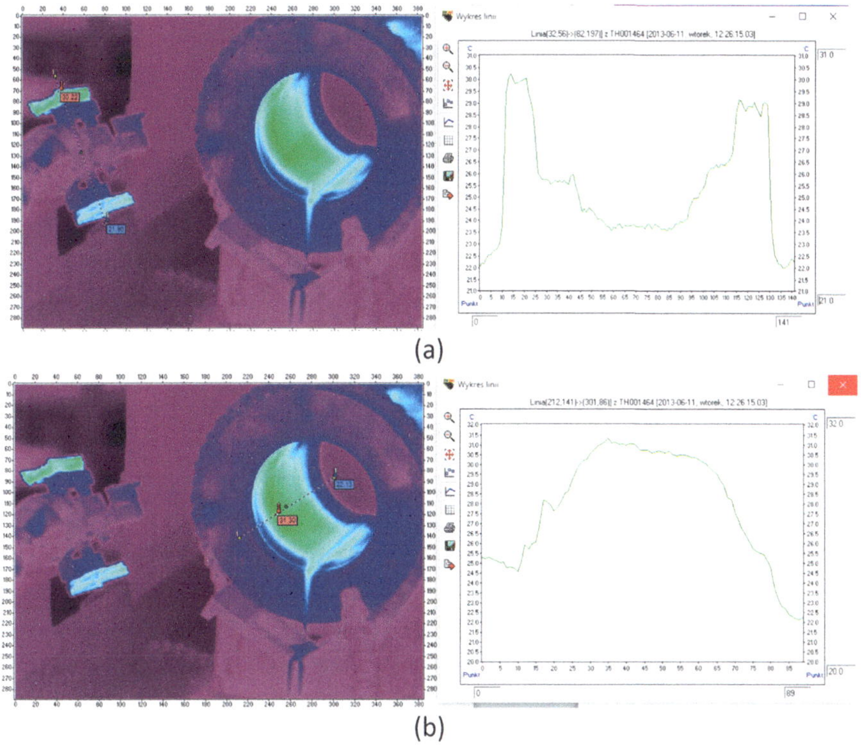

Fig. 2. Temperatures generated in honing process: (a) temperature of oilstone T_o = 21,96 °C; (b) temperature of honed cylinder liner T_w = 31,3 °C.

2 Methodology

Figure 3 presents the unrolled surface of the honed holes, on which a quadrangular area with its vertices points: (0, 0), (n, n), (n, − n), (2n, 0) was separated. Figure 3 also shows a straight line y = c dividing showed area into two halves.

The grain track is shown in a thick line, running in any direction in the axial and radial direction.

The greater angle inclination of the path V_1 with respect to the radial direction axis means the higher axial velocity or the lower amount of honing head rotation, the smaller inclination angle V_2 with respect to the radial direction axis means lower axial velocity or higher amount of rotation of honing head (Fig. 3).

Abrasive grain at time $t = 0$ (Fig. 3) is in point (0, 0). At times $t = 0, 1, 2 \ldots 2n$ abrasive grain in the honing process performs "jumps". Each "jump" consists in shifting the grain trajectory point by one unit to the right and at the same time by one unit down or up. At time $t = 2n$, if the grain were at the point (2n, 0), it would mean that the same number of upward and downward shifts were made. Each broken line connecting the possible position of the points on which the abrasive grains moved at times $t = 0, 1, 2 \ldots, 2n$ is called the trace of the abrasive grain. There are possibilities to

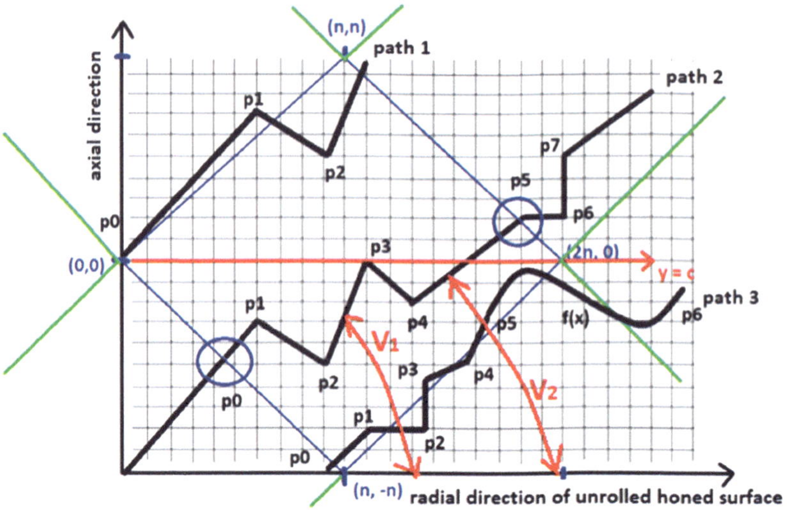

Fig. 3. Sample abrasive grain paths on unrolled Surface of honed holes.

find the number of possible traces of the moving grain and the number of traces without common points in the straight line with the equation y = c (c - the natural number, c < n). All traces of the point are found inside or on the perimeter of the square with the vertices: (0, 0), (n, n), (2n, 0), (n, n).

During the abrasive grain straight movement there are n jumps up and n jumps down direction. The number of possible traces θ is equal to the number of n-element combinations from the set with 2n elements, i.e. equal to θ = (2n/n).

The formula on the curve curvature radius *R* of a *Z(X)* abrasive grain trajectory move functions is given as (1):

$$R = \frac{\left[1 + \left(\frac{d}{dX} Z(X)\right)^2\right]^{\frac{3}{2}}}{\left|\frac{d^2}{dX^2} Z(X)\right|} \quad (1)$$

The formula on the length of a *Z(X)* abrasive grain trajectory move functions from point *a* to point *b* on circumference's axis *X* is given as (2):

$$L = \int_a^b \sqrt{1 + \left[\frac{dZ}{dX}\right]^2} dX \quad (2)$$

The curvature values of the curve *Z(X)* of the abrasive grain trajectory will be presented in the paper as a number equal to the number of revolutions of the honing head occurring for one travel of the honing head in the upper direction during honing with variable kinematic conditions (Fig. 4).

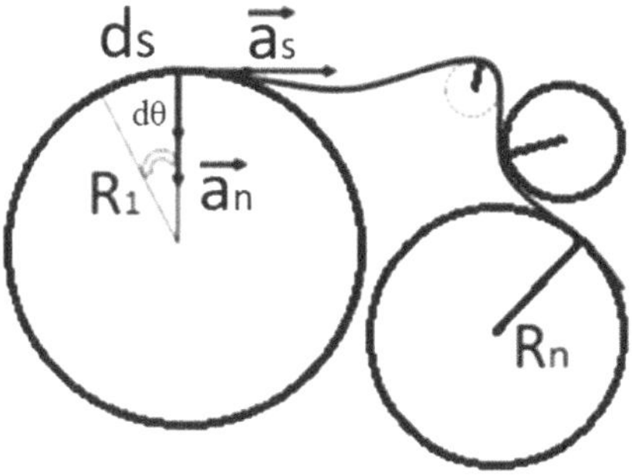

Fig. 4. The curvature of honed oil paths: $R = ds/d\theta$ – the curve curvature radius; $ds/d\theta$ – increase of the grain path move relative to the angle increment, a_n – normal acceleration, a_s – tangential acceleration.

The formula on the speed of abrasive grain movement in motion with acceleration $a = 0$ (3):

$$V = \frac{L}{t} \tag{3}$$

where: L – path length [m],

t – time [s].

The average speed of grain movement in motion with acceleration $a \neq 0$ is given by the formula (4):

$$V_{av} = \frac{L_2 - L_1}{t_2 - t_1} = \frac{dL}{dt} \tag{4}$$

where: L_2 - end path length,

L_1 - beginning path length, t_2 - end time, t_1 - beginning time.

Total acceleration a is the acceleration of the tangential and normal acceleration (5):

$$a = \sqrt{a_n^2 + a_s^2} \tag{5}$$

Tangent acceleration $\overrightarrow{|a_s|}$ is characterized by the rate of change of the numerical value of the speed of abrasive grain movement (6):

$$\overrightarrow{|a_s|} = \frac{dV}{dt} \tag{6}$$

Normal acceleration $\overrightarrow{|a_n|}$ is characterized by the speed of change of grain speed move direction (7):

$$\overrightarrow{|a_n|} = \frac{V^2}{R} \tag{7}$$

Variable kinematic of the honing process is the best way of honing of cylindrical holes especially with a thin wall of honed parts. Figure 5 is showing the abrasive grain trajectory received during honing with variable rotations 100–200 min^{-1} of honing head in up and down direction. The graph showed on Fig. 5 presents the trajectory of the abrasive grain on the developed surface of the honed hole for one complete honing cycle lasting t = 1,7 s and containing 5 revolutions of honing head. This way of presenting of honed grain path trajectories help to receive the function *Z(X)* of a single abrasive grain trajectory in honing with the variable kinematic condition.

The graph shown on Fig. 5 presents the function of the fourth degree polynomial, which is not able to properly adjust to the trajectory of the abrasive grain moving up and down during variable honing condition, for this reason the next graphs will present fragments of the abrasive grain trajectory described using the function of the third degree polynomial moving during the honing only upwards. The number of calculated steps depends on the number of rotations of honing head for each graph.

Figures 6, 7, 8, 9, 10 and 11 presents the number of radiuses of the curve curvature R of honing abrasive grain trajectories curves Z(X) received in variable kinematic condition with different parameters of the honing process.

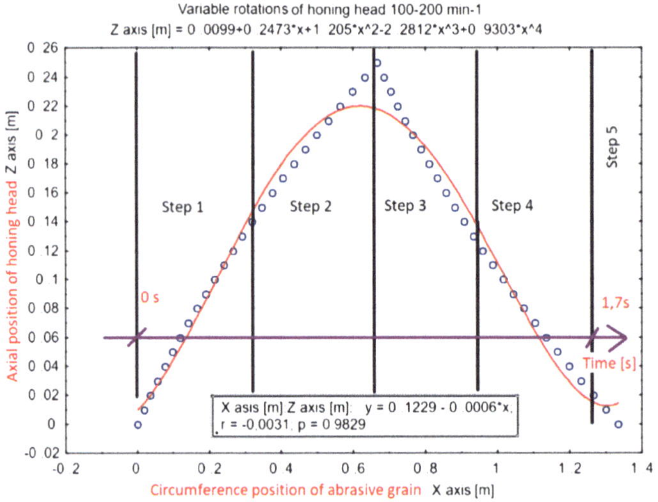

Fig. 5. Variable rotations of honing head 100–200 min^{-1}, up and down move, 5 steps (5 turns of honing head) S1, S2…S5 per 1 machining cycle lasting 1,7 s.

$$Z(X) = 0{,}0079 + 1{,}4522x - 4{,}0292x^2 + 5{,}3009x^3$$

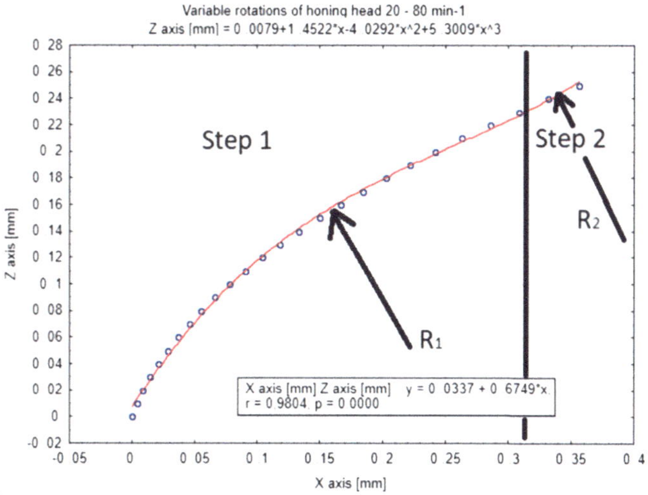

a)

b) $R_1 = 0{,}504$ m, $R_2 = 0{,}589$ m ; c) $L_1 = 0{,}392$ m, $L_2 = 0{,}048$ m, $L_1 + L_2 = 0{,}44$ m,
d) $a_{n1} = 0{,}29$ m/s^2, $a_{n2} = 2{,}109$ m/s^2

Fig. 6. Variable rotations of honing head 20–80 min^{-1}, up move direction of honing head; temperature increase of honed workpiece in 1 min of honing is 3.025 °C; grain trajectory length on the circumference length X = 0.356 m, total acceleration a = 0,37 m/s^2; (b) curve curvature radius in the middle of each honing cycle step; (c) path length in each honing steps, (d) normal acceleration amount in each honing step.

Figure 12 presents a research stand that enables smooth possibilities to change all parameters of the process during honing.

Figure 13 presents the influence of the sum of curve curvatures radius ΣR of the function's $Z(X)$ on temperatures of honed workpiece T increase (a) and the influence of the normal acceleration value a_n on the honed workpieces temperatures T (b). The amount of sum of curve curvatures radius ΣR of abrasive grain trajectory path $Z(X)$ length 0,438 m and the sum of the amount of normal acceleration a_n = 2,4 m/s^2 had the lowest impact on honed workpieces temperature.

Figure 14 presents the influence of the length of abrasive grain trajectories L on workpiece temperature T. The length of the abrasive grain path L from 0,53 m had the lowest impact on honed workpieces temperature T.

$$Z(X) = 0{,}0022 + 1{,}1849x - 2{,}2866x^2 + 2{,}5942x^3$$

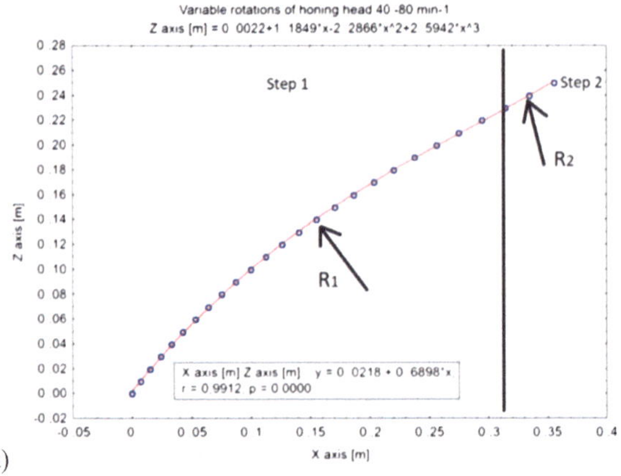

a)

b) $R_1 = 0{,}807$ m, $R_2 = 0{,}765$ m; c) $L_1 = 0{,}391$ m, $L_2 = 0{,}047$ m, $L_1 + L_2 = 0{,}438$ m,
d) $a_{n1} = 0{,}193$ m/s^2, $a_{n2} = 1{,}095$ m/s^2

Fig. 7. Variable rotations of honing head 40–80 min^{-1}, up move direction of honing head (a); temperature increase of honed workpiece in 1 min of honing is 5,03 °C; grain trajectory length on the circumference length X = 0,356 m, total acceleration a = 0,246 m/s^2; (b) curve curvature radius in each honing step; (c) path length in each honing; (d) normal acceleration amount in each honing step.

$$Z(X) = 0{,}0002 + 0{,}9169x - 0{,}8464x^2 + 0{,}6817x^3$$

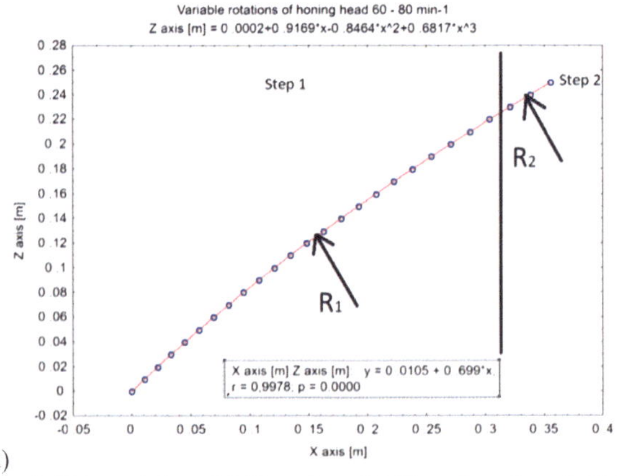

a)

b) $R_1 = 1{,}735$ m, $R_2 = 1{,}476$ m, c) $L_1 = 0{,}388$ m, $L_2 = 0{,}048$ m, $L_1 + L_2 = 0{,}436$ m,
d) $a_{n1} = 0{,}095$ m/s^2, $a_{n2} = 0{,}177$ m/s^2

Fig. 8. Variable rotations of honing head 60–80 min^{-1}, up move direction of honing head (a); temperature increase of honed workpiece in 1 min of honing is 5.75 °C; grain trajectory length on the circumference length X = 0,3560474 m, total acceleration a = 0,123 m/s^2; (b) curve curvature radius in each honing step; (c) path length in each honing; (d) normal acceleration amount in each honing step.

$$Z(X) = 0{,}69774 + 0{,}5583x - 0{,}2172x^2 + 0{,}0906x^3$$

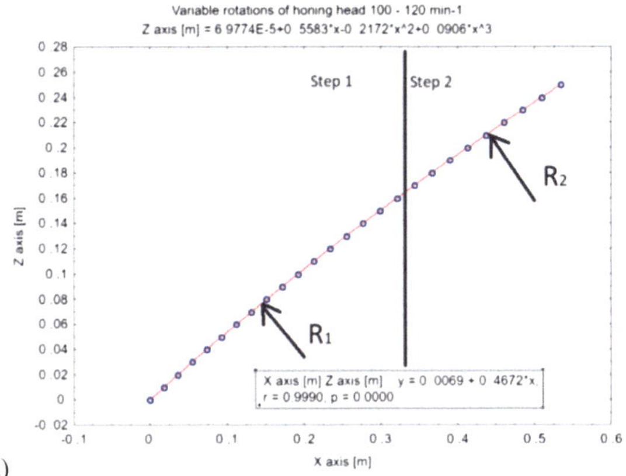

a)

b) $R_1 = 3{,}989$ m, $R_2 = 3{,}793$ m; c) $L_1 = 0{,}351$ m, $L_2 = 0{,}239$ m, $L_1 + L_2 = 0{,}59$ m,
d) $a_{n1} = 0{,}086$ m/s^2, $a_{n2} = 0{,}199$ m/s^2

Fig. 9. Variable rotations of honing head 100–120 min^{-1}, up move direction of honing head (a); temperature increase of honed workpiece in 1 min of honing is 5.216 °C; grain trajectory length on the circumference length X = 0.534 m; total acceleration a = 0,123 m/s^2; (b) curve curvature radius in each honing step; (c) path length in each honing; (d) normal acceleration amount in each honing step.

$$Z(X) = 0{,}0004 + 0{,}5446x - 0{,}3333x^2 + 0{,}1653x^3$$

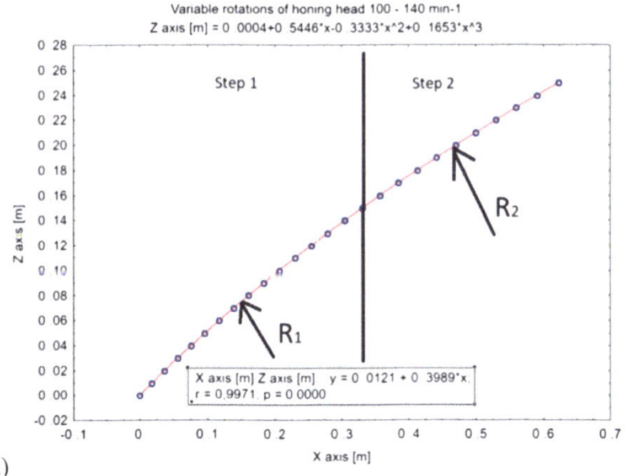

a)

b) $R_1 = 2{,}588$ m, $R_2 = 2{,}578$ m, c) $L_1 = 0{,}346$ m, $L_2 = 0{,}332$ m, $L_1 + L_2 = 0{,}678$ m,
d) $a_{n1} = 0{,}582$ m/s2, $a_{n2} = 1{,}647$ m/s^2

Fig. 10. Variable rotations of honing head 100–140 min^{-1}, up move of honing head (a); temperature increase of honed workpiece in 1 min of honing is 7,524 °C; grain trajectory length on the circumference length X = 0,62308301 m; total acceleration a = 0,246 m/s^2; (b) curve curvature radius in each honing step; (c) path length in each honing; (d) normal acceleration amount in each honing step.

$$Z(X) = 0,0006 + 0,5347x - 0,3616x^2 + 0,1816x^3$$

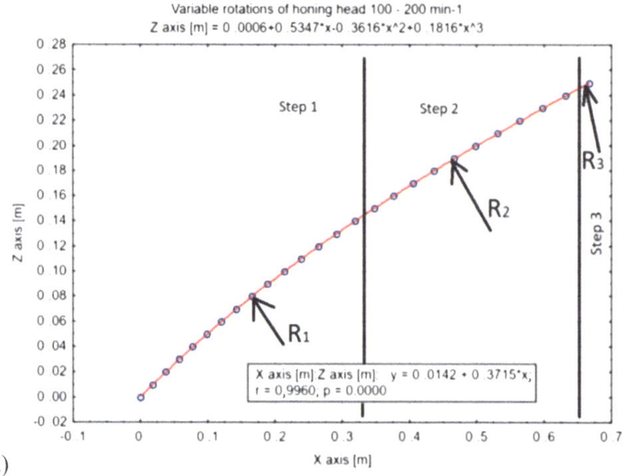

a)

b) $R_1 = 2{,}348$ m, $R_2 = 2{,}348$ m, $R_3 = 2{,}042$ m; c) $L_1 = 0{,}343$ m, $L_2 = 0{,}33$ m, $L_3 = 0{,}041$ m, $L_1 + L_2 + L_3 = 0{,}714$ m; d) $a_{n1} = 0{,}7$ m/s^2, $a_{n2} = 1{,}928$ m/s^2, $a_{n3} = 3{,}32$ m/s^2

Fig. 11. Variable rotations of honing head 100–200 min^{-1}, up move of honing head: (a) grain trajectories, (b) curve curvature radius in each honing step; temperature increase in 1 min of honing: 22,701 °C; grain trajectory length on the circumference length X = 0,667588 m; total acceleration a = 0,616 m/s^2; (b) curve curvature radius in each honing step; (c) path length in each honing; (d) normal acceleration amount in each honing step.

(a) (b)

Fig. 12. Research stand: (a) conventional hydraulic honing machines WMW SZS-200; (b) sample photo of honed surface with grain trajectories received in variable kinematic conditions of honing process.

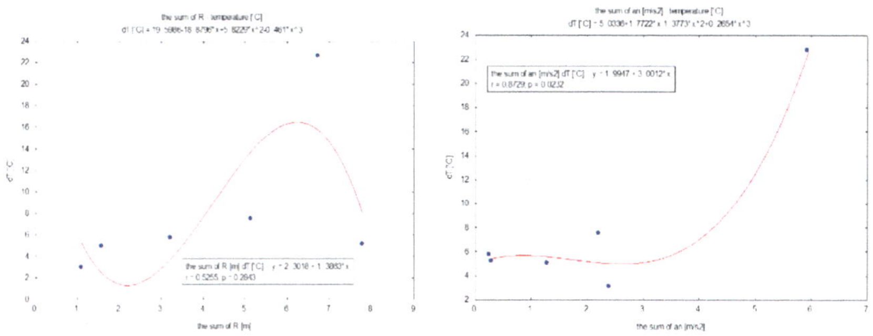

(a) (b)

Fig. 13. (a) The influence of the sum of curve curvatures radius ΣR on honed workpiece's temperatures; (b) the influence of the amount of normal acceleration a_n on honed workpiece's temperatures.

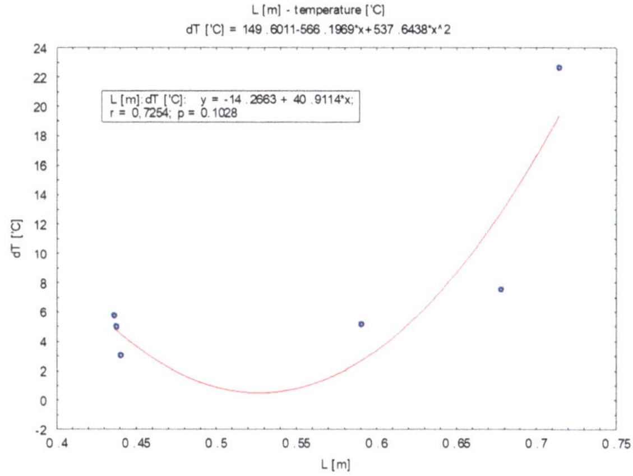

Fig. 14. The influence of the length of abrasive grain trajectories L on honed workpiece's temperatures T.

Figure 15 presents the influence of the sum of curve curvatures radius ΣR and abrasive grain path length L on honed workpiece temperature T. The lower ΣR amount affects the honed workpiece temperature T decrease for a longer path length L.

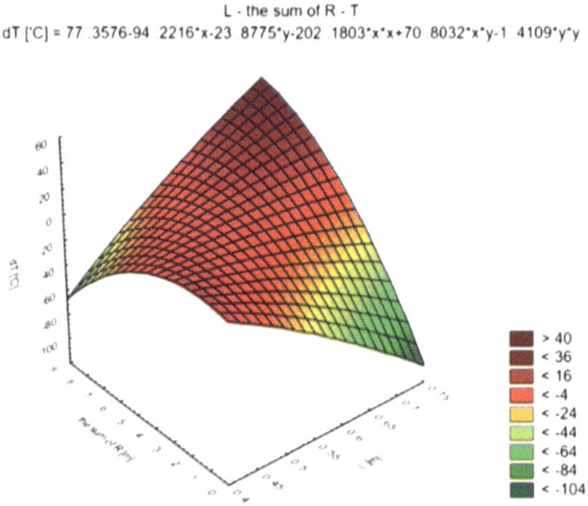

Fig. 15. The influence of the sum of curve curvatures radius ΣR and abrasive grain path length L on honed workpiece's temperature T.

Figure 16 shows an example of curves with the same path length and with different amplitudes and curvatures. The choice of the desired shape of the trajectory can be supported by the analysis of the radii of curvature of abrasive grain moves function $Z(X)$.

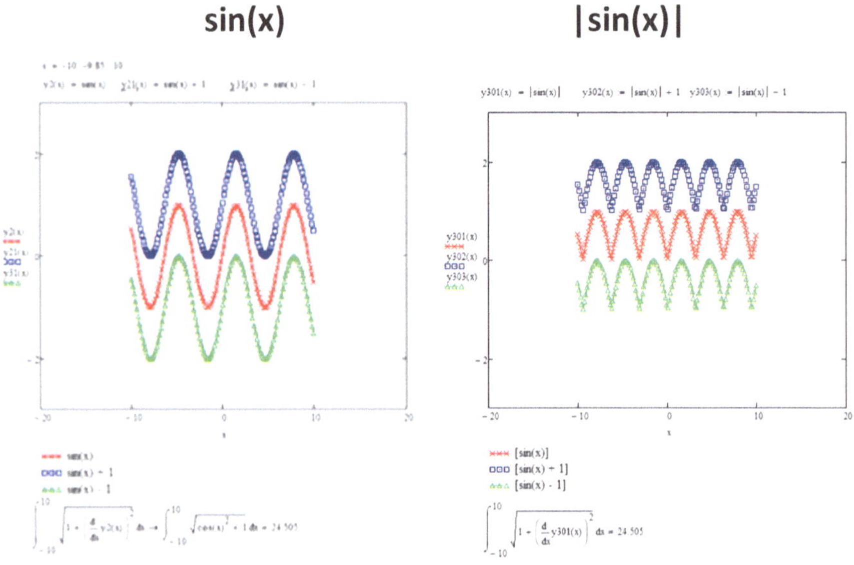

Fig. 16. An example of abrasive grain trajectories $Z(X)$ curves, with the same path length and with different amplitudes and curvatures.

3 Conclusions

The lower value of the sum of curve curvatures radius ΣR, of abrasive grain trajectories *Z(X)* function, affects the honed workpiece temperature *T* decrease for a longer path length *L*.

The curve curvature radius *R* of the abrasive grain paths function *Z(X)*, in the variable kinematic condition of honing of cylindrical holes, can be used to improve honing of thin wall cylinder liners, due to decreasing of amount temperature received in honing process.

Detailed knowledge of the impact of the curvature of the abrasive grain path on the machining process can bring many benefits in the development of methods of honing of cylindrical holes, especially in the case of thin-walled workpieces.

Further research should make it possible to verify the legitimacy and benefits of introducing into the control of modern CNC honing machines the possibility of programming of the trajectory of the abrasive grain, using a function describing the motion of the honing head.

References

1. Miller, J.: Shape curve analysis using curvature. A dissertation submitted to the University of Glasgow, Department of Statistics (2009)
2. Barylski, A., Sender, P.: Experimental studies of diameter and temperature increments of workpieces during honing the long holes under production conditions. Mechanik **9**, 34–43 (2014). (in Polish)
3. Buj-Corral, I., Vivancos-Calvet, J., Coba-Salcedo, M.: Modelling of surface finish and material removal rate in rough honing. Precis. Eng. **38**, 100–108 (2014)
4. Sender, P.: Variable kinematics of honing process - influence on machined workpiece. In: Proceedings of the Conference on Trends in the Development of Machinery and Associated Technology TMT 2018, Karlovy Vary, Czech Republic, pp. 1–4 (2018)
5. Sender, P.: Variation of technological processes of machining. In: Szantyr J. (eds.) Modern Technologies and Energy Conversion, vol. II. Faculty of Mechanical Engineering at the Gdansk University of Technology (2013)
6. Sender, P.: The influence of the abrasive grain trajectory on the machining of thin-walled cylinder liners of internal combustion engines with variable kinematics of honing. Autobuses **19**(12), 634–641 (2018). (in Polish)

Monitoring System with a Vision Smart Sensor

Anna Zawada-Tomkiewicz(✉) and Dariusz Tomkiewicz

Koszalin University of Koszalin, Koszalin, Poland
anna.zawada-tomkiewicz@tu.koszalin.pl

Abstract. Current changes in the manufacturing techniques, referred to as Industry 4.0, require developing them as cyber-physical systems. And such a cyber physical approach has been used in the elaboration of the vision smart sensor (VSS) applied for the evaluation of machined surface during the cutting process. The VSS was built of several functional elements. The functional integration is based on the 5C architecture – Connection, Conversion, Cyber, Cognition and Configuration Levels. The starting point of the vision smart sensor is the machine vision system. Its components and connections with other levels are presented in a functional setup in the Smart Connection Level. The whole data processing and estimation of image features are described in the Data-to-information Conversion Level and the Cyber Level. The whole architecture of the cyber physical production system is preceded by discussion on elements within the Cognition Level. The vision smart sensor consists of spatially integrated components. The core of the sensor is the data processing unit, based on a PC architecture, to control the data processing, storing the data and communication. Image acquisition control and other auxiliary devices are controlled with this software. The paper concludes with the current research challenges required to expand the application of a cyber physical approach in the vision smart sensor design to machined surface monitoring. As an example, the vision smart sensor was implemented for the surface roughness estimation while cutting hardened steel with PCBN tools.

Keywords: Cyber Physical System · Machining · Machined surface · Vision smart sensor

1 Introduction

The constantly growing demand for precision manufacturing entails a requirement to apply a cyber-physical approach for production process design [1, 2]. The requirements for the system are to identify factors affecting machined surface, to make all relevant production data available, to analyze these data, to extract information from these data, and finally to generate knowledge about the current state of the machined surface from the information [5, 7].

The cyber physical approach in developing the production system expands the cutting process by applying the monitoring of the process and the machined surface [6, 11]. Figure 1 presents a cyber physical production system in which the quality of a machined surface is defined when the process is in a design stage. The metrological model of the surface for further stages of manufacturing is developed by mapping the

M. Majewski and W. Kacalak (Eds.): IIRTS 2019, LNME, pp. 94–105, 2020.
https://doi.org/10.1007/978-3-030-37566-9_9

requirements of the surface. The metrological model in the next stage is transformed into the manufacturing model. The physical realization of cutting process according to the manufacturing model is the final stage in the creation of the machined surface.

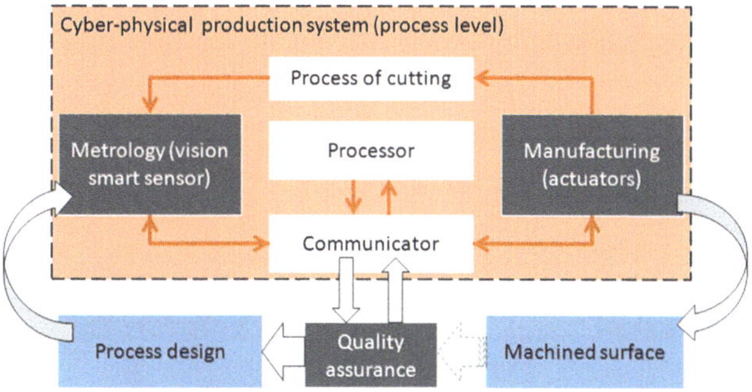

Fig. 1. Cyber-physical approach in precision manufacturing of machined surface.

In the cyber physical production system, the preparation of the machined surface means that the surface fulfils the demanded functionality. It is ensured by the functional and spatial integration of the measurement system and its implementation in the real world. The aim of applying cyber-physical approach is to elaborate the measurement system, which enables not only the data transfer but also providing the necessary information for the Cyber, Cognition and Configuration Level [6, 11].

The functional integration of the vision smart sensor is presented in the application of the machined surface monitoring tasks. The functional integration is discussed in Sect. 2 on the basis of the 5C architecture. The levels are characterized separately and then there is the whole system architecture discussed. Section 3, in turn, relates to the spatial integration of the system where all the components of the vision smart sensor are implemented in the cyber physical production system.

2 Cyber-Physical Model of the Vision Smart Sensor

The aim of the vision smart sensor development was focused on elaborating a low-cost device able to measure and evaluate the machined surface quality during the cutting operation.

The vision smart sensor was designed as a combination of the several subsystems, which were functionally connected into one system: camera and sensors, data integration and image analyzer, control of optics and illumination, actuators for optics and illumination (Fig. 2). Camera and sensor block is responsible for gathering data from environment - especially vision data about the state of the monitored surface. The data from the camera sensor block were processed in the data processing block. The data processing block was divided on two functional subsystems. The first subsystem was responsible for data integration and image analyses. The second subsystem was responsible for the control of actuators. All subsystems constituting vision smart sensor exchange data throughout machine to machine interface (M2M). The M2M consists of data transmission protocols such as: Ethernet, IEEE1394, RS232 and other. The M2M is also responsible for communication with additional systems, which support proper operation of the vision smart sensor. These additional systems embrace: the camera fastening subsystem, the work zone cleaning subsystem, human machine interface and other subsystems communicating through machine-to-machine interface.

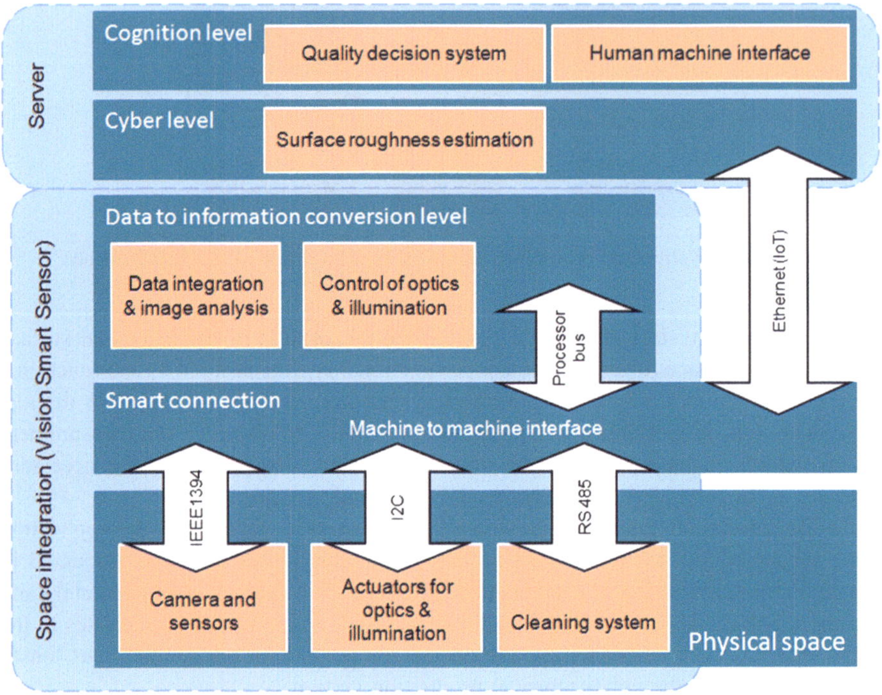

Fig. 2. The architecture of the vision smart sensor.

2.1 Smart Connection Level

The attributes of the Smart Connection Level can embrace such elements as Plug & Play, wireless communication and sensor networks [4, 12]. The elaborated system, in physical space, consists of a surface illumination system, a magnifying system, a frame grabber with an image pre-processing system. The output of the first functional component is a surface image with strictly defined quality.

The surface illumination system is based on white LEDs constituting directional lighting. The system is equipped with an actuator able to change the position of the light source. The input position is the elevation angle of 30°. The lighting, together with optical system, is centered in relation to the axis of the nominal workpiece.

The magnifying system is used to obtain a digital image featuring suitable field of view and depth of view. With the field of view of 1 mm^2 and depth of view of 100 μm, the magnifying is appropriate for the surface monitoring where cut-off length is recommended as 0.8 mm. The magnifying system is composed of lenses and extender sets.

The frame grabber and image pre-processing are connected due to the demand for high quality of image information. First, the application of a digital camera with IEEE 1394 interface for image acquisition provides the possibility of on-line control of the gathered data. Secondly, the single frame was collected and converted into a two-dimensional array of 8-bit numbers. Finally, the process of image pre-processing was performed to test image quality. Only images of sufficient quality were used for further processing [10].

2.2 Data-to-Information Conversion Level

The second level of the Cyber Physical System refers to such attributes as prediction of performance and degradation of system components, smart analytics for quality of data and multi-dimensional data correlation.

The second functional component of the vision smart sensor, the image processing system, consists of algorithms which provide the option to transform image data into feature space. Surface image data are acquired with the use of a camera and an optical system. The optimization of the acquisition system is crucial for the further stages (Figs. 3 and 4 "Surface image acquisition"). Once data are obtained, the image enhancement methods are applied and mathematical parameters from the data are calculated. In this way, the first step embraces the set of predestined methods of image description, such as statistical space and spectrum features, wavelet features and other. All the features are then examined with the pruning method and only six normalized features are selected to be used for further processing (Figs. 3 and 4 "Surface image processing").

The selection of features is performed in the training mode. After the first image is acquired and its quality is satisfactory, four groups of features are analyzed with respect to their mutual correlations and their dispersion. On this basis, the best six features are selected from each group. Then the correlations between the selected features and the target model of the surface parameters are investigated with the pruning method. The selected features are then used at the Cyber Level.

2.3 Cyber Level

At the Cyber Level of a cyber physical production system, the models for components are elaborated basing on the previously acquired experimental data and with application of data mining algorithms for clustering [8, 9]. The third functional component, decision making process, is strictly connected with the goal of monitoring. It consists of a classifier, which can correlate image features with actual process parameters [3].

For the estimation of process parameters, the recurrent neural network estimator was applied. After the initial process of preparing the architecture of the network, the feed-forward three-layer neural network was determined. The Extended Kalman Filter was used as the algorithm for neural network training. This algorithm enables optimal estimation of state variables in the presence of noise. The advantage of this algorithm is its recurrent method of working. As with the neural network estimator the training and decision making modes are applied.

The training mode of the vision smart sensor was presented in Fig. 3. For a cutting process being part of the cyber physical production system the process design (Fig. 1) embraces the model of the machined surface and the measurement plan. According to these elements the proper surface texture parameters are determined. Simultaneously, the parameters of the surface formation model in machining are identified. This model is used to generate model values of surface texture parameters useful for calibrating the estimator in a decision making mode.

In the measurement plan the process and the machined surface are evaluated using the vision smart sensor. As shown in Fig. 3, apart from other layouts (surface cleaning or lighting correction), the system has also a module that checks the quality of a single image. This is an important issue because in the decision making mode the quality of the surface is evaluated using changes in the surface image. Any disturbance in the image may generate a false signal that the parameter assessed for the process and surface has changed significantly with adequate consequences.

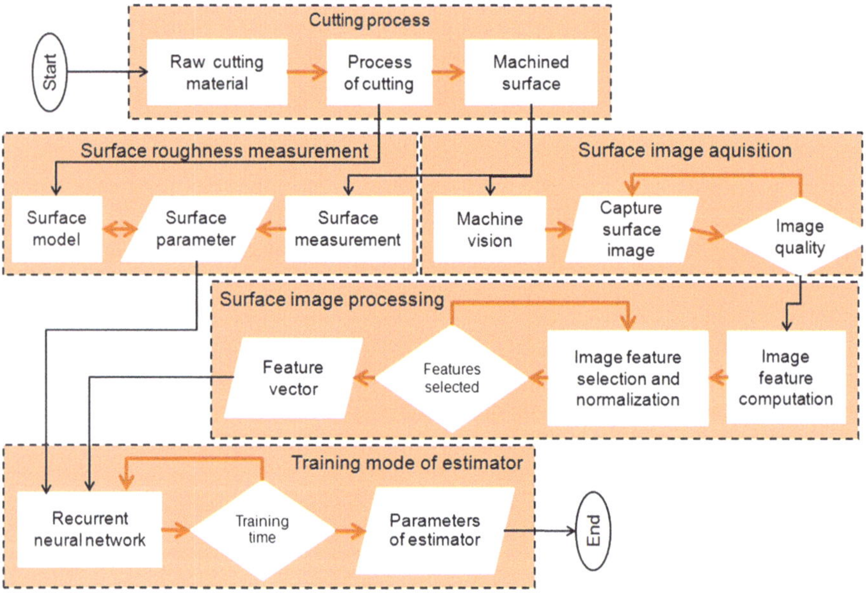

Fig. 3. The training mode of the vision smart sensor.

Then the surface image is processed within which the raw data are pre-processed and the image features are calculated. As was stated earlier the architecture of the neural network estimator is elaborated to work in a recurrent way. The neural network estimator calculated the value of a parameter in the next step on a base of the estimated value of a surface roughness parameter in the current step and the increments of image features.

Estimation of process parameters follows changes in a surface image and relays the warning when the changes are above a certain criterion. Any parameter correlated with surface roughness can be used. The estimated value is a nonlinear function of time, where the value of the surface parameter in the next step depends on the value of the parameter in the current step and the increments of surface image features.

The image features have been calculated, and their increments have been used in the estimation. The increments of image features were used to incorporate the current changes in the machined surface image into the estimator. The neural network weight vector was adjusted after several presentations of training set so that the network error was below the threshold. The process of training was then stopped.

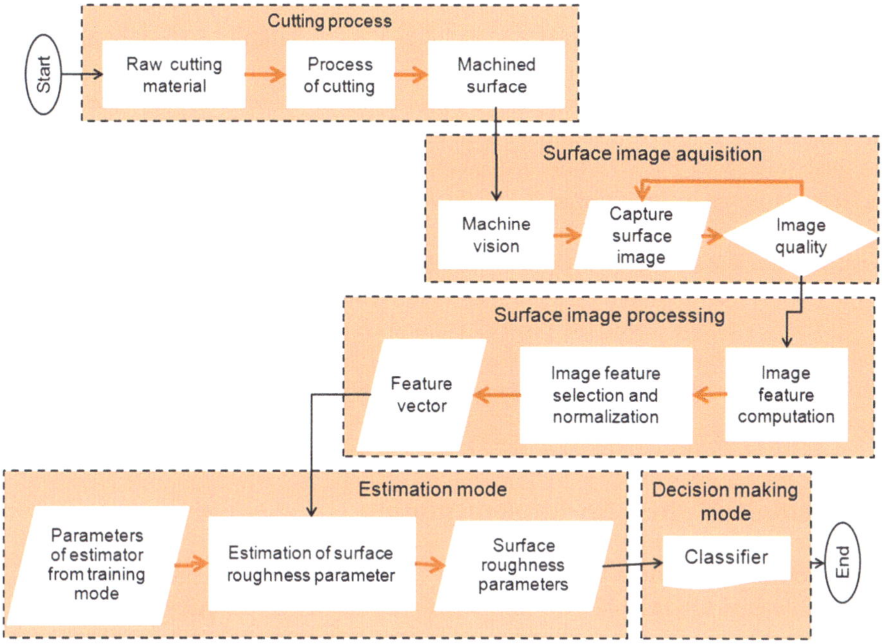

Fig. 4. The decision-making process flow for surface roughness monitoring.

One of the advantages when using the recurrent neural network in a form of the Kalman filter was the possibility of on-line estimation of model parameters. Such an approach enabled the surface monitoring system to be adjusted interactively by the system (or operator) during the process of cutting. When the estimation error achieved the threshold value, the algorithm changed its mode from training to parameter estimation (decision-making mode). The estimator parameter database was created for various cutting parameters and materials so that the time of adjustment of monitoring system was as short as possible. The estimation algorithm was composed of four stages (Fig. 4):

- Image acquisition with the use of the machine vision system (Connection Level),
- Image feature determination (Data-to-Information Conversion Level),
- Training of the recurrent neural network on data from a new cutting process (parameters calculated on a base of the surface formation model from a data base). It was done by the optimization of weight values in such a way that the difference between the estimated value and measured value was below the defined threshold (Cyber Level and Cognition Level – training mode),
- Surface roughness parameter estimation. It was done by the parameter estimation via the recurrent neural network with optimized weights for defined cutting parameters (Cyber Level and Cognition Level – decision making mode).

2.4 Cognition Level

Cognition Level provides an integrated simulation and synthesis, decision-making and remote visualization for operator both for training and decision making modes.

Model of the system at the Cognition Level relates to the overall goal of the system - assurance of machined surface quality in a cutting operation. Defined in such a way, the goal influences the structure and system behavior. As the quality of machined surface can be predicted only in a limited range, the surface data obtained from the Cyber Level must be directed for decision making. The estimated surface parameters must be sufficient to evaluate (1) their current values, (2) their relation to upper and lower confidence level, and (3) their statistical characteristics.

At the Cognition Level, the process parameters are visualized for the operator to evaluate the relation of the actual value of defined surface parameters to the reference values. At the level, all the potential uses of the surface with its usability in assembly are considered.

3 Example of the VSS Implementation

The vision smart sensor is based on surface image data from machine vision [9, 10]. The main element – machine vision system – can be used for image acquisition and pre-processing. The demanded quality of image data imposes a requirement for the optical system and illumination system to be controlled. Thus, the complexity of the machine vision system increases and constitutes the basis for the vision smart sensor design. The maximum estimation time was assessed to be less than one second with the accuracy suitable for surface monitoring.

3.1 System Operation

The vision smart sensor was mounted on a CNC machine tool in such a way so that the working zone could be safely and efficiently observed. (Figure 5). The vision smart sensor was elaborated for an easy adaptation considering different work material, different tool and conditions. In each case the vision smart sensor was shied from external factors such as heat and chips.

The vision smart sensor was elaborated as a monitoring system. The core of the sensor is the **data processing unit**, based on PC architecture. The system was elaborated to control the processing of the data, also storing and communication. Image acquisition and other auxiliary devices control are supported with this software. The LabView environment was used, which enables the implementation of complex and diversified tasks. The digital Basler A311fc camera was connected to data transmission unit through the interface IEEE 1394. The set of functions from the IMAQ Vision library was used for the image acquisition and their further pre-processing.

The communication of the VSS with the machine tool software for the goal of the system calibration and the surface roughness estimation is embedded in the soft sensor software. Ethernet and other interfaces are used to transmit data between the vision smart sensor and other elements of the control system.

The main goal of the human machine interface (HMI) was to provide the human operator with information about the process. The human machine interface fulfilled three main functions in the system. The first function performed by HMI was to provide information for the machine vision system calibration. The second HMI function is used in the process of training a neural network. The third HMI function can be defined as a general supervision of the process. HMI provided the ability to switch between the training and decision making mode. This function was necessary due to the periodical calibration of the estimator and enabled the tuning of the vision smart sensor for the new cutting conditions. The human operator, who could decide to change the process parameters, performed all the activities concerning the human machine interface.

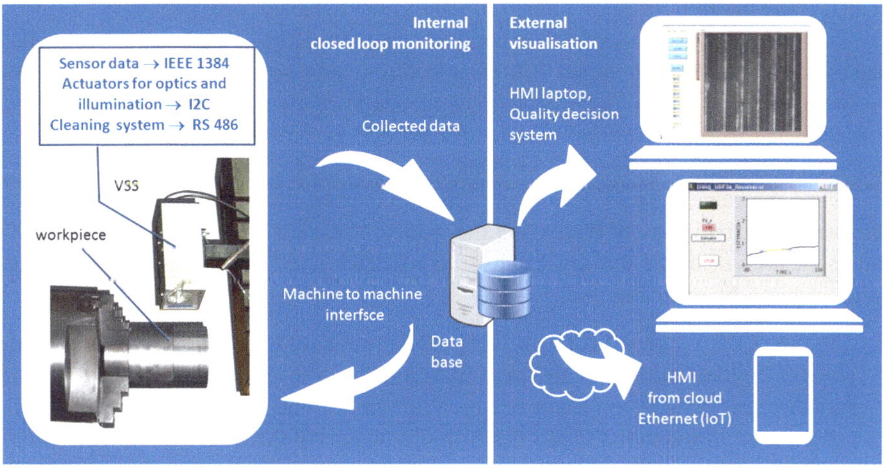

Fig. 5. The vision smart sensor mounted on CNC machine tool.

After the initialization and adaptation of the machine vision system (camera, lenses, illumination, and cleaning system) there is a possibility for the system to get the surface image and evaluate its quality. The parameters of the camera can be adjusted during its use and the brightness, gain and speed of the shutter are subjected to a process of optimization. HMI has been designed to display the on-line image and, upon the user's request, to download a single frame. Regardless of the operator's assessment the system can operate in the automatic mode.

If the neural network estimator does not require training, the system goes directly into the decision making mode. The first step of the estimation relates to the initial value of surface roughness parameter and also the increments of the image features. The initial value of surface roughness parameter in the first step is taken from the surface formation model developed based on previous data (data downloaded from the database). For the first step, it should be assumed that the increments are equal to zero. Due to the need to determine increments in the next steps, a single image from the image stream is collected, for which the values of image features are determined. In the next steps, the estimated value is determined from the estimated value in the current step and the increments of the image's feature values.

If the neural network estimator needs to be trained, it is necessary to provide the surface parameters for this process (activities connected with the optimization of the neural network weights). This means that additional data (surface parameter values) must be provided for the entry of the neural network. Experience shows that the training process is short-lived and after a few steps the system goes from training mode to decision making mode. Switching the system does not reset the parameters of the estimator. The estimator in the decision making mode estimates the value of the surface parameter in the next step based on the estimated value in the current step and the increase in the value of the image's features. It is possible to adjust the value of the parameters of estimator accordingly to the measured value at any time. In the conducted tests, the frequency of saving all data in the database was 1 s.

Additionally, the system software was elaborated in such a way that the estimated values of parameters were tested with the use of the moving average. If the differences in subsequent moving averages exceed the threshold of the controlled value, an alarm was triggered. If the controlled value exceeded the set threshold, then the system was designed to initiate an alarm. The software was developed to listens to the data channel and when the value appears, it displays it in a numerical form and as the next point in the chart. The frequency of displaying the value of estimated parameter was 2 s. With the application of the vision smart sensor any surface roughness parameter or process parameter can be estimated.

- The use of the Extended Kalman Filter can be limited to the decision making mode when the initial parameters are taken from the database (Fig. 4). It is the fastest and simplest use of the vision smart sensor. Without calibration, the vision smart sensor can be used to monitor the process or roughly control the quality of the machined surface.
- The Extended Kalman Filter can be calibrated with the use of training (Fig. 3). In a training mode the selection of image features can be performed (long training) or the system can be tuned to the specified cutting conditions (short training). In each case of training mode, it is necessary to know the reference values of the selected parameter. After the training process is finished a recurrent neural network switches to the decision making mode (Fig. 4).

3.2 Results and Discussion

The vision smart sensor was implemented for the surface roughness *Ra* parameter. The turning dry tests were carried out on NEF 400 machine tool for cutting speed 165 m/min, feed rate 0.15 mm/rev and a depth of cut equal to 0.1 mm. Workpiece material was EN 41Cr4 steel, hardened to 40 HRC. The tests were performed with ISO TNGA 160408 S01020 Sandvik PCBN inserts CB7020 with thin 1 µm TiN coating - Ti (75%) and N (25%). Grains have a diameter of 1 µm to 2 µm. The cutting inserts were placed in the DTGNR 2525 M 16 tool holder.

The vision smart sensor was developed to process image data of the machined surface. Mathematically, a machined surface image is defined by one matrix describing the variability of shades of gray. Taking into account the character of the machined surface image and its basic pattern, the representation which was very suitable for the goal of image analysis was wavelet decomposition. The digital wavelet transform was performed for 100 cross-sections of the image, 0.8 mm in length. Coiflet filters were used to decompose the original cross-section of the image into seven components: the approximation at 6th decomposition level (*A*6) and the details at 1st ÷ 6th decomposition level (*D*1 ÷ *D*6). The portions of data were described with the statistical measures. With the use of Optimal Brain Surgeon Method the data set was reduced to the six: *SD* (standard deviation) of *A*6, *MD* (maximal difference) of A6, *LP*3 (Minkowski norm) of *D*5, *NAE* (normalized absolute error) of *D*3, *LP*3 of *D*3, *NAE* of *D*2.

For a given cutting test of hardened steel, a reference cutting process was carried out and the *Ra* parameter of the surface roughness resulting from the measurements was introduced as the reference value of the neural network estimator. The architecture of the neural network was optimized. Seven neurons were applied in the input layer, three neurons in the hidden layer (with the activation function in a form of hyperbolic tangent) and one neuron in the output layer (with the linear activation function). The Extended Kalman Filter was applied for training the neural network.

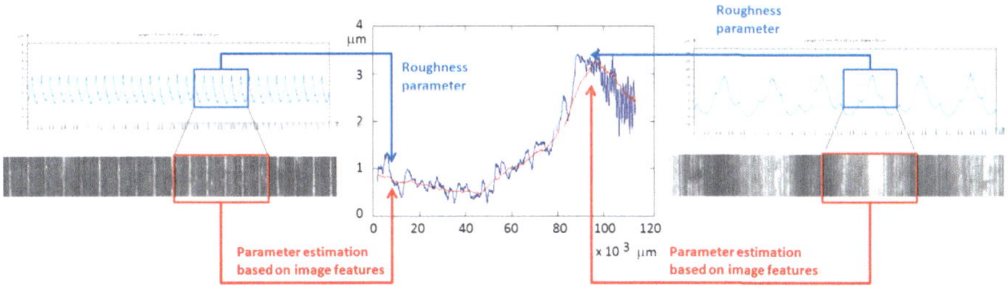

Fig. 6. An example of *Ra* surface roughness estimation by a recurrent neural network.

The vector of weights was taken from the database, and then modified according to the reference values with the use of the recurrent algorithm. Short-term training of the neural network was used. After several presentations of the reference value, the vector of weights was optimized so that the results of both estimated and reference values overlapped. Then the vision smart sensor changed its mode from training to decision making mode. The neural network has begun estimating the value of surface roughness parameter. The predicted value was estimated based on a current value and the differences in surface image features. After several iterations, the system changed from the training mode to the decision making mode. The results of the estimation are presented in Fig. 6 (red line).

After the cutting process was finished, the surface roughness values were measured along the workpiece and *Ra* values were determined (Fig. 6 - blue line). The chart shows a large correspondence between the estimated value and the measured value. What's more, the process of training was short. Comparing the estimated values (red line) and the reference values (measurement of surface roughness after the cutting process) the estimated values are less dispersed around the moving average. This is due to the features of Extended Kalman Filter, which can perform efficiently even in the presence of noise.

In the example, the cutting process was elaborated so that the changes related to occurrence of vibrations appeared during the cutting process and could be identified on the machined surface. The estimation of the *Ra* parameter was estimated correctly also in this case. The values of the estimated parameters were similar in statistical sense to those that resulted from the measurements.

4 Conclusions

The purpose of the presented research was to design a vision smart sensor with the use of cyber physical approach. This approach allowed for the separation of functions in the 5C system and the adaptation of the physical elements of the system for the best spatial integration.

- In the 5C system at the lowest level of the Connection Level, a communication system was developed between the elements. This level includes the method of data collection, data storage and transmission.
- At the Data-to-Information Conversion Level the data are transformed into the meaningful, which are correlated with the purpose of the production system. In the case of the vision smart sensor, the acquired image data are converted into image features. At this level, the normalization of features and their selection are also performed.
- At the Cyber Level, the input data are converted into an estimated value of surface roughness parameter. In the case of the vision smart sensor, a recursive algorithm was used. The Extended Kalman Filter algorithm predicts the estimated value for the next step using the image features and the current estimated value. Estimated data can be on the measure, weight or class level.
- At the Cognition Level, decisions are made to use estimated data to optimize the manufacturing process.

The physical implementation of the vision smart sensor was based on the fact that the machine vision system was embedded in the machined surface monitoring system. As part of spatial integration, the human machine interface was also developed. As an example of the veracity of the vision smart sensor operation, cutting tests were carried out for hardened steel and the vision smart sensor was used to estimate the roughness parameter *Ra* of the machined surface. The vision smart sensor was able to systematically estimate the value of the parameter, display it and save it in the database. After the process was finished, the surface was measured using Talysurf CLI 2000. Then the *Ra* parameter was determined as a function of the cutting path. The consistency was observed between the estimated and the measured values.

References

1. Ng, A.H.C.: Virtual manufacturing for press line monitoring and diagnostic. Int. J. Mach. Tools Manuf. **48**, 565–575 (2017)
2. Lipiński, D., Kacalak, W.: Metrological aspects of abrasive tool active surface topography evaluation. Metrol. Meas. Syst. **23**(4), 567–577 (2016)
3. Mannan, M.A., et al.: Application of image and sound analysis techniques to monitor the condition of cutting tools. Pattern Recogn. Lett. **21**, 969–979 (2000)
4. Min, C., et al.: Machine-to-machine communications: architectures, standards and applications. KSII Trans. Internet Inf. Syst. **6**(2), 480–497 (2012)
5. Neugebauer, R., et al.: Mechatronic systems for machine tool. Annals of the CIRP **56**(2), 657–686 (2007)
6. Rzevski, G.: On conceptual design of intelligent mechatronic systems. Mechatronics **13**, 1029–1044 (2003)
7. Simmon, E. et al.: A vision of cyber physical cloud computing for smart networked systems. NISTIR, 7951 (2013)
8. Wu, Q., et al.: Theoretical and experimental investigation of spindle axial drift and its effect on surface topography in ultra-precision diamond turning. Int. J. Mach. Tools Manuf. **116**, 107–113 (2017)
9. Zawada-Tomkiewicz, A.: Estimation of surface roughness parameter based on machined surface image. Metrol. Meas. Syst. **17**(3), 493–504 (2010)
10. Zawada-Tomkiewicz, A., Storch, B.: Classifying the wear of turning tools with neural networks. J. Mater. Process. Technol. **109**(3), 300–304 (2001)
11. Zawada-Tomkiewicz, A., Wierucka, I.: A case study in technological quality assurance of a metric screw thread. Measurement **114**, 208–217 (2018)
12. Zawada-Tomkiewicz, A., Tomkiewicz, D.: Mechatronic monitoring tool (MMT) for machined surface evaluation during the cutting process. In: Proceedings of 11th IMEKO TC14 Symposium on Laser Metrology for Precision Measurement and Inspection in Industry, LMPMI, pp. 373–376 (2014)

Analysis of the Internal Load the Ankle Joint Module as the Basic Structural Assembly of the Lower Limb Rehabilitation Exoskeleton

Szczepan Śpiewak[1(✉)], Jan Awrejcewicz[2], and Wojciech Kunikowski[2]

[1] Faculty of Mechanical Engineering and Computer Science, Czestochowa University of Technology, Częstochowa, Poland
szczepan_spiewak@poczta.onet.pl
[2] Department of Automation Biomechanics and Mechatronics, Lodz University of Technology, Łódź, Poland
awrejcew@p.lodz.pl, wojciech.kunikowski@gmail.com

Abstract. The publication presents a global outline of the construction of the lower limb rehabilitation exoskeleton. The main focus was put on the structure of the exoskeleton's ankle assembly cooperating with the patient's ankle joint. Based on the geometrical properties of the assembly a three-dimensional model was developed using the Finite Element Method (FEM) for determining the stresses and displacements of components subjected to variable loads. In the FEM model, finite elements of the 3D-solid type and beam type were used. To determine the external loads an experiment was performed using a dynamometric platform by means which the ground reaction forces, excited by a patient wearing the exoskeleton, were measured. During testing and analysis, particular attention was focused to orthotropic ABS material components manufactured with 3D printing technology. The work contains partial data obtained from strength tests of samples manufactured by a 3D printer. The results of measured strength parameters that describe the mentioned orthotropy are implemented by defining a suitable material model in the FEM model of the tested assembly. As a result of the numerical calculations performed, the stress fields of individual elements of the assembly were obtained. The distribution of stresses depends on the relative position of patient's foot relative to the ground, on the loads generated by driving modules and inertial forces. Based on these stress values their amplitude variation were determined which enabled selecting parts that met the requirements construction strength, durability and safety of use of rehabilitation exoskeleton.

Keywords: Exoskeleton · Quasi-static gait walking · Strength analysis · Finite Element Method (FEM) · Material model of Akrylonitryl-*co*-butadien-*co*-styren

1 Introduction

Taking into account the experience in human rehabilitation and in the restoration of movement abilities of lower limbs presented in [1, 2], it can be stated that it is particularly important to start this process as early as possible. In many cases,

M. Majewski and W. Kacalak (Eds.): IIRTS 2019, LNME, pp. 106–117, 2020.
https://doi.org/10.1007/978-3-030-37566-9_10

rehabilitation is reduced to the cyclic movement of limbs excited by work of the physiotherapist. One of the devices, which in some stages of rehabilitation can significantly assist the rehabilitation work is an exoskeleton.

The publication [3] summarizes ten different design exoskeleton used for rehabilitation or in order to increase the motor performance of human knee joint. One example of an exoskeleton which ensures the stability of movement by means of adequate limb fastening system and an integrated motor drive system is presented in [4].

In order to increase the availability of exoskeletons for the greatest possible number of people in need of rehabilitation, the manufacturing costs of such a system should minimised. It can be ensured by: use of cheap and easy accessible materials and semi-finished products, simplified connections between the modules responsible for assisting muscles in the joints, introduction of compensating elements that approximate the trajectory of human joints with the pure rotation, the use of unified drive systems which are easily controlled, the use of 3D printing based rapid prototyping techniques of complex shape components which are otherwise hard or time consuming to manufacture.

The entirety of assumptions is particularly important both in design and prototype phase. The proposed design solution of lower limb rehabilitation exoskeleton based on mentioned assumptions is described in Sect. 2.

In order to ensure the safe use of the exoskeleton not only the drive system and its control units, but also the strength requirements of individual parts must be take into account. The durability of exoskeleton's main elements will be determined by the internal loads caused by the impact of the drive units and forces of inertia of the patient. Ranges of forces generated by human muscles during the movement per unit of patient's weight are described in [5, 6]. Strength calculations require creating a numerical model that takes into account geometrical parameters of the analysed system, interactions between different parts of the exoskeleton and the boundary conditions, including external forces, identification of effective degrees of freedom and their accurate reproduction.

Determination of stress values using classic analytical equations is effective especially for simple shaped machine parts made with isotropic and homogeneous materials. However, in the case of more complex geometry of individual elements of the proposed system and particularly high amount of contact points, more effective is the use of numerical methods such as FEM. In order to identify stresses that may appear in the elements of the designed exoskeleton, a FEM model was developed, the structure of which is described in Chapter 3. The literature provides examples of models used in strength calculations of exoskeleton elements. For example, in [7] an approximated dimensional optimization of exoskeleton elements including muscles within the knee joint is presented. Optimization criterion targeted the weight reduction of parts. The publication [8] includes an FEM analysis of the internal load conditions of the selected components of the exoskeleton in the static loading condition. Presented FEM models can be considered as very simplified models, because they do not include, among others, interactions between the components (contact conditions). Obtained stress contours in individual parts of the exoskeleton with defined load conditions representing boundary states during human gait allow for identification of types of fatigue cycles. Identifying the nature of the load will simplify the estimation of the durability of elements - measured by the number of cycles.

2 Exoskeleton Prototype

Based on the above assumptions, a prototype of a rehabilitation exoskeleton was developed and manufactured. Its structural model and photographs (along with a patient during the tests) are presented in Fig. 1a and b.

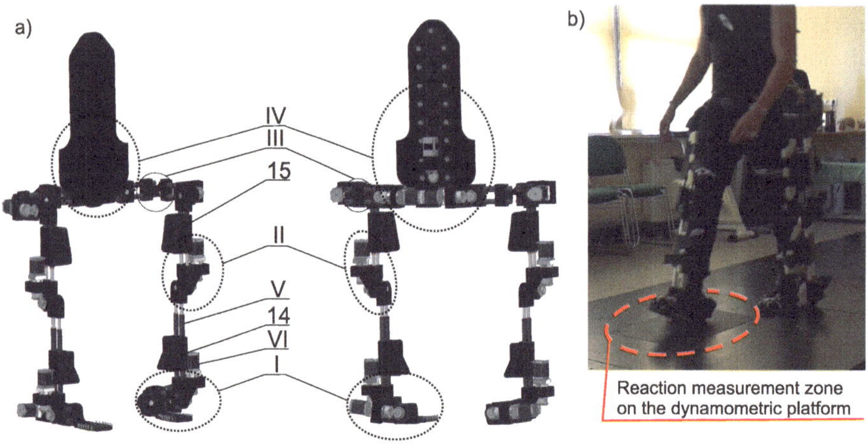

Fig. 1. Visualization of the exoskeleton prototype (a) and measurement of the ground reaction force (performed by dynamometric platform) during human gait (b).

The exoskeleton consists of six main modules, whose task is to excite movement within the human musculoskeletal system. Modules indicated in Fig. 1a are: I - ankle joint assembly, II - knee joint assembly, III – hip joint assembly, IV - spine support structure, V - elastic connectors, VI - drive module. Elements that connect the patient's limbs to the exoskeleton are thigh mount (14) and shank mount (15).

The basic module of the exoskeleton presented in this publication is the ankle joint assembly, which detailed construction is shown in Fig. 2a. The drive module (electric motor with planetary gearbox and conical gear wheel) is located between the engine clamp (1) and the engine mount (2). The set of elastic connectors is clamped through the casing of the outer sleeve of the elastic connector (3). Elements 1, 2, 3 are connected together by use screws (4). The flexion of the foot is accomplished by rotating the shaft (12) connected via a pin (13) to regulation clamp (5). Connection between parts (5) and heel module connectors (6) serves also as a means for adjustment of the dimensions to user's foot. Inversion and eversion motion of the foot is realized via the shaft (11) coupled by a bolt (10) with heel module (7). Heel module connectors are fixed to heel module connectors mount (9). For free movement of patient's toes the metatarsal module is responsible (8). It consists of a number of links adjustable to the size of the patient's foot. The elastic connectors are responsible for the coupling of assemblies I-II and II-III. In addition, they perform the function of compensating trajectory of human joints to simple rotation. In FEM models, these elements were assigned to the knee joint model.

3 FEM Model of the Exoskeleton Ankle Joint Assembly

In order to limit the computational time of the FEM analysis, individual exoskeleton assemblies created separate models with appropriately defined degrees of freedom. The same calculations for the entire system had to be carried out in a defined order starting from the ankle assembly model. The obtained values of reaction forces in finite elements located at the dividing planes between parts of the exoskeleton (i.e. the edges) become the active forces for the FEM analysis of the next assembly. This is dictated by the condition of maintaining the continuity of the internal forces in the division planes of the geometrical structure of the exoskeleton's modules.

Finite element mesh model of an ankle joint module (Fig. 2a) is presented in Fig. 2b.

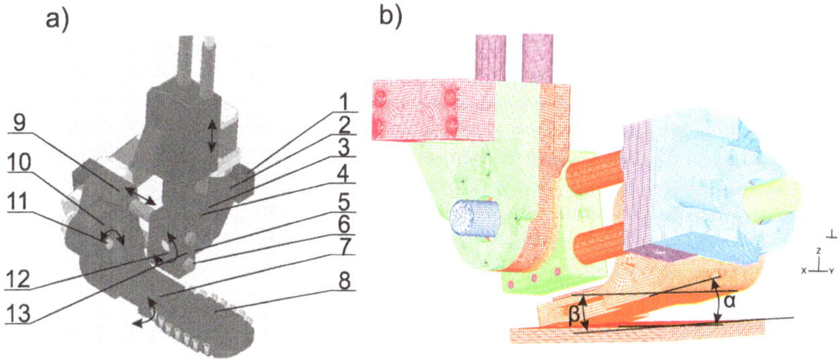

Fig. 2. Construction (a) and mesh of the FEM model (b) of the ankle joint module.

In order to identify material parameters and mutual relations between parts of the assembly I (Fig. 2 a), 25 groups of finite elements were defined whose mutual positions shown in Fig. 3. These groups represent the geometrical structure of, respectively: EGI-1 first engine clamp, EGI-2 first engine mount, EGI-3 external sleeve of elastic connector, EGI-4 housing of the outer sleeve of elastic connector, EGI-5 bearing, EGI-6 cover, connecting screws EGI-7, EGI-8 conical heads of screws, EGI-9 first drive shaft, EGI-10 another stet of screws with EGI-11 their heads, EGI-12 regulation clamp, EGI-13 pins, EGI-14 heel module connectors, EGI-15 second engine mount, EGI-16 heel module connectors mounts, EGI-17 second drive shaft, EGI-18 second bearing, EGI-19 second cover, EGI-20 next group of screws and their EGI-21 heads, EGI-22 bolt, EGI-23 patient's foot, EGI-24 ankle module and EGI-25 ground.

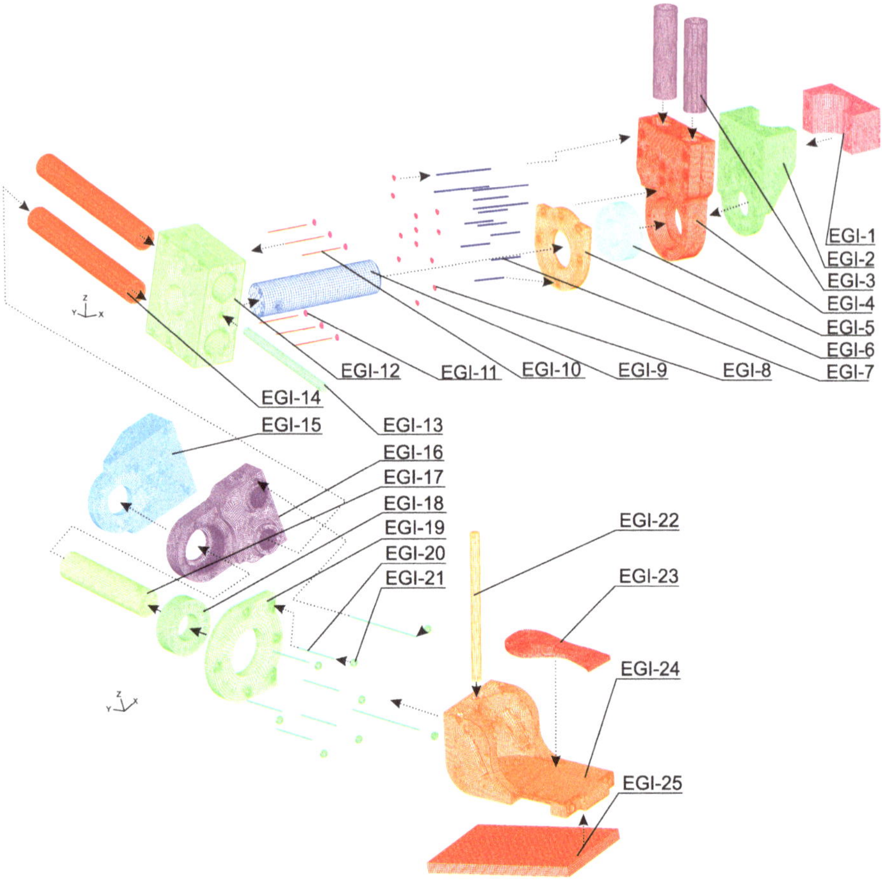

Fig. 3. Groups of finite elements of an ankle joint assembly.

Most groups are described by eight-node finite elements of the 3D-Solid type [9, 10]. Exceptions are groups representing screws: EGI-7, EGI-10, EGI-20. Similarly to the work [11], the bolts were modelled in a simplified way with two-nodal finite elements type a beam [9, 10]. For such elements material characteristics, cross-section shape and dimensions and the value of the bolt's preload are defined. After performing the calculations the values of axial forces acting on the bolt can be acquired. As mentioned in the FEM model, it is possible to define material properties for individual groups of finite elements. The prototype used materials such as aluminium (used for the drive shafts), steel (used for screws, bearings and pins and to determine the properties of a rigid ground surface), polypropylene (used to make the outer sleeve of the elastic connector) and other components made by 3D printing method from the ABS material. In the FEM model, linear-elastic isotropic material models were used with the defined Young's modulus and Poisson's ratio for aluminium E = 7000 MPa, ν = 0.35, for steel E = 200,000 MPa, ν = 0.33, for polypropylene E = 2000 MPa, ν = 0.3 and for tissue E = 21500 MPa, ν = 0.3. Since the printed out component of ABS material exhibits orthotropy resulting from applying the following layers in the forming process [14] the internal forces will depend on the orientation of the application of loads relative to the main plane of the printer. For each group of finite elements made from ABS material an *ab* plane compatible with the printer plane was defined. In order to find material parameters a static stretching and torsion tests for cylindrical samples were performed. The samples were made in two orientations (vertical and horizontal) with respect to the main plane of the print *ab*. In the process of printing, layers are applied to the *ab* plane along the *c*-axis. As a result of performed strength tests the following values of the orthotropic model parameters were obtained: E_a = 1026 MPa, $E_b = E_c$ = 1115 MPa ν_{ba} = 0.49, $\nu_{cb} = \nu_{ca}$ = 0.2, G_{ba} = 405 MPa, $G_{cb} = G_{ca}$ = 425 MPa. In order to model the interactions between screw head surfaces and conical sockets, the surface contact conditions have been defined including penetration and friction, with the coefficient of friction μ = 0.1 [9, 10, 12]. Contact conditions of this type have been introduced for pairs of interacting surfaces, described by nodes of appropriate finite element groups, i.e.: EGI-1 - EGI-2, EGI-2 - EGI-4, EGI-3 - EGI-4, EGI-4 - EGI-5, EGI-4 - EGI-6, EGI-5 - EGI-6, EGI-4 - EGI-8, EGI-6 - EGI-8, EGI-11 - EGI-12, EGI-12 - EGI-14, EGI -12 - EGI-13, EGI-14 - EGI-16, EGI-9 - EGI-13, EGI-15 - EGI-16, EGI-16 - EGI-18, EGI-16 - EGI-19, EGI-16 - EGI-21, EGI-19 - EGI-21, EGI-22 - EGI-24, EGI-17 - EGI-22, EGI-24 - EGI-25. It should be taken into account that the coefficient of friction between EGI-23 - EGI-24 groups will depend on whether the patient's foot will be covered with fabric. Locations of the EGI-23 group nodes were obtained by performing a laser scan of the patient's footprint of size EUR 42. Displacements of the extreme nodes of the beam elements were determined by means of ideally rigid links [9, 10] connected to all nodes located on the frontal surfaces of the conical heads and the abutment surfaces representing the nuts. The rolling bearings in the model were replaced by the rings (EGI-5 and EGI-18) with external dimensions equal to those of mentioned parts. Between cylindrical surfaces, common to nodes from EGI-5 - EGI-9 and EGI-17 - EGI-28 groups, assumed contact conditions were similar to those described above but taking into account an appropriate for rolling bearings friction coefficient μ = 0.0015 [13]. In Fig. 8 the angles α and β are determined, whose initial values can be parametrically changed in the model. This is required to simulate the internal loads of the assembly depending on the position of the patient's foot relative to the ground. The positions of nodes representing individual components of the assembly were parametrically related to the angles α and β by introducing local coordinate systems. These systems are defined relative to the global coordinate system by determining the coordinates of the starting point and the Euler angles. The coordinates of the nodes from groups from EGI-9 to EGI-16 were assigned to the first local coordinate system, while finite nodes from EGI-17 to EGI-24 were assigned to the second coordinate system. Nodes from the EGI-25 group is assigned to the third local coordinate system.

Basic boundary conditions of the ankle assembly model are presented in Fig. 4.

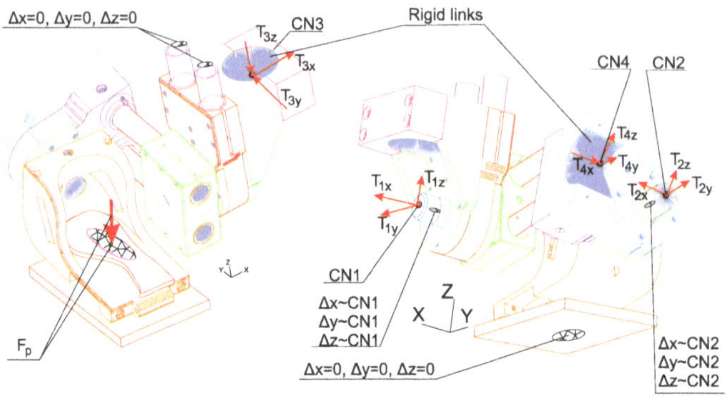

Fig. 4. Boundary conditions of the ankle assembly model.

Zero displacement values were defined for nodes of group EGI-25 localised on the lower surface that in the model is mapping of the ground and for the nodes of group EGI-3 containing the dividing plane between assemblies I and II of the exoskeleton structure. The resultant force of foot pressure F_p on the heel module was spread over the external nodes of the EGI-23 group. The forces resulting from the operation of the driving system were replaced with concentrated moments. Values of those moments were calculated based on the forces component between teeth of conical gears with a pitch cone angle $\theta_1 = \theta_2 = 45°$ and nominal moment M_n.

$$T_{3y} = F_o \cdot \left(\frac{tg\alpha_n \cdot sin\theta_1}{cos\beta_n} + tg\beta_n \cdot cos\theta_1 \right) \cdot \frac{d_m}{2} + F_o \cdot \left(\frac{tg\alpha_n \cdot sin\theta_2}{cos\beta_n} - tg\beta_n \cdot cos\theta_2 \right) \cdot a_1 \quad (1)$$

$$T_{3x} = F_o \cdot a_1 \quad (2)$$

$$T_{3y} = F_o \cdot \left(\frac{tg\alpha_n \cdot sin\theta_2}{cos\beta_n} + tg\beta_n \cdot cos\theta_2 \right) \cdot \frac{d_m}{2} + F_o \cdot \left(\frac{tg\alpha_n \cdot sin\theta_1}{cos\beta_n} - tg\beta_n \cdot cos\theta_1 \right) \cdot a_2 \quad (3)$$

$$T_{1x} = F_o \cdot a_2 \quad (4)$$

Where:

d_m- average pitch diameter of the conical gear wheel
F_o - circumferential force equal to $2M_n/d_m$
α_n - nominal pressure angle of the conical gear wheel
β_n - helix angle of the conical gear wheel
a_1, a_2 – distances of the point of attachment of the circumferential force vector from the support (bearing) of the drive shaft and driven shaft in the drive train.

The vectors of concentrated moments were attached in additionally defined central nodes marked in Fig. 4 as CN1, CN2, CN3, and CN4. Location of nodes CN3 and CN4 is related to the centre of gravity of the hole formed by the association of the engine clamp and the engine mount (Fig. 2a). Moments located in these two nodes represent the driving side while the driven side is modelled by the nodes CN1 and CN2, located in the centre of gravity of the driven gears. The interaction of clustered moments in central nodes on dependent nodes (representing the according parts of the assembly) was implemented by introducing perfectly rigid links [9, 10]. CN1 and CN2 nodes were coupled the nodes located on the external surface of the shaft (Fig. 2a) with the length equal to the length of the key under the toothed wheel. This the dependence results from the necessity of recreating the twisting moments. In view of the geometrical identity of the groups EGI-2 and EGI-15 values of the other moments introduced to the model can be determined by assuming that $M_n = T_3 = T_{1x} = T_{4x} = T_{2y}$, $T_{3y} = T_{4z}$, $T_{3x} = T_{4y}$, $T_{1y} = T_{2z}$, $T_{1z} = T_{2x}$.

4 Strain Simulation of Assembly I

In the strength analysis of parts that are periodically affected by time varying loads in order to determine their durability it is necessary to know the value of various threshold stresses. The course of stress variation allows characterizing the type of cycle and amplitude. Therefore, it was decided that based on FEM model, stress distributions for characteristic setting position of heel module relative to the ground with customized external loads (under the boundary conditions) should be calculated. It was performed basing on a quasi-static analysis. The first calculations were made for the so-called initial conditions in which the elements of the assembly are loaded only with the screws pre load forces (predetermined value S = 1000 N). The screw tension is defined in beam-like finite elements (with geometric parameters of M5 screws). The next calculations were made in accordance with the conditions of the FEM model described above (Fig. 4) at zero values of the angles between the heel module and the ground ($\alpha = \beta = 0°$) - position 1 (Fig. 5).

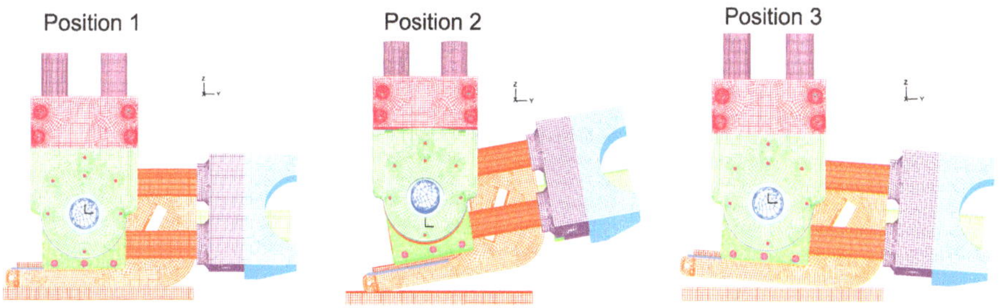

Fig. 5. Positions of finite element groups of ankle assembly corresponding to the extreme positions during human gait.

The parameters of the model for position 1 simulate the start of the lifting off the foot from the ground forced by the drive system (with the assumed force $F_p = 450$ N and the value of the driving moment $M_n = 40$ Nm). In the case of the position 2 the initial values of angles are: $\alpha = 10°$. $B = 5°$. Changes in the angles α, β (Figs. 2b and 5) in the model are achieved by the transformation of the local coordinate systems in which, mentioned in the previous chapter groups of finite elements were defined. The position 3 recreates the moment of impact of the heel on the ground. The amplitude of the load at position 3 was determined experimentally using a dynamometric platform (Fig. 1b) with the participation of the patient outfitted with the exoskeleton (total weight 90 kg). The time series of values of the vertical component of ground reaction force is shown in Fig. 6a.

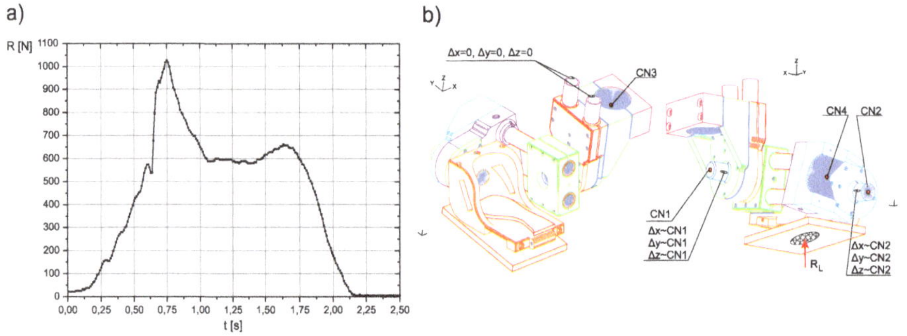

Fig. 6. Time series (a) of ground reaction force and new boundary conditions (b) for FEM model in position 3.

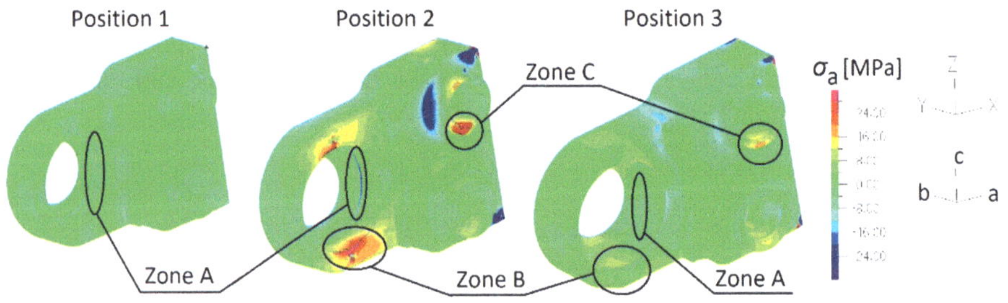

Fig. 7. Distribution of normal stresses σ_a for different positions of the model of the heel module connectors' mount (finite element group EGI-16).

Maximum value of reaction R (time $t = 0.75$ s) corresponds to the maximum overload in normal use of the exoskeleton when the total pressure carried by one of the heel modules. In order to perform calculations for the indicated position with regard to the reaction force, new boundary conditions of the model were defined, as illustrated in Fig. 6b. Reaction R of the modified model was defined as an active load evenly distributed on the lower nodes of group EGI-25. The central nodes CN1, CN2, CN3,

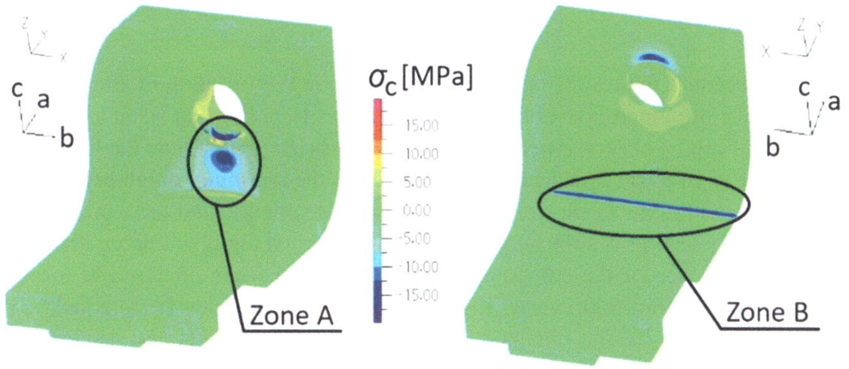

Fig. 8. Distribution of normal stresses σ_c for position 3 of the model of the heel module (finite element group EGI-23).

and CN4 lost the possibility of rotation. In this way, the holding torque in the drive module is recreated. Rest of the boundary conditions have not been changed.

After the calculations were performed, the fields of normal stresses were determined (denoted as σ_a, σ_b, σ_c) and related to the orthotropic structure of the material model and the field of reduced stresses referred to the isotropic material models. In the assessment of the safety of use of a given component made of the isotropic material being tested, the yield limits R_e with values of: 35 MPa for polypropylene, 240 MPa for aluminium and 300 MPa for steel were taken into account. These values were exceeded during the calculations in finite element model groups EGI-14 and EGI-17. In contrast, the minimum values of strength determined experimentally based on a static tensile test of samples with orthotropic structure according to defined directions are R_{ma}= 26 MPa, R_{mb}= R_{mc}= 20 MPa. These values have been exceeded for the parts of group of finite elements: EGI-4, EGI-12.

Method for conducting analysis of the results obtained in the field of fatigues strength of the orthotropic components properties may be presented by example of the heel module (Fig. 7) and the heel module connectors' mount (Fig. 8).

In the case of the heel module connectors' mount it was found that the maximum stresses are close to the threshold stress in the direction a. In the case of the heel module, it is the direction c. Considering the stress fields shown in Fig. 7 it can be concluded that the variability of the stress on the zone A can be described by a non-symmetrical cycle as limit stress value at a different positions of the assembly changes sign ($\sigma_{Amin} = -24$ MPa, $\sigma_{Amax} = 16$ MPa, which gives an amplitude of $\sigma_{Aa} = 20$ MPa). In the case of zones B and C, the variation in stress can be described by a one-sided repeated cycle with an amplitude $\sigma_{Ba} = \sigma_{Ca} = 24$ MPa. Similar condition applies to stresses in the heel module (Fig. 8) in zone A covering area of a pivot connection and the zone B as the first zone to contact (linear) with the ground.

5 Conclusions

Taking into account the results of the FEM analysis on the variation of stresses during human gait for presented assembly parts with respect to performed static strength and fatigue tests it can be observed that the boundary stress are greater than the fatigue limit and imply the high-cycle resistance range. However, the minimum durability measured by the number of fatigue cycles for the obtained stress variation and the work related to biocompatibility testing is sufficient. For example, the first signs of fatigue that can be found on the surface of the component under inspection when analysed in the work load should occur after 500 cycles (the amplitude of the repeated cycle equals σ_{Ba} = $_{Ca}$ = 24 MPa - Fig. 7). The obtained distributions of stress fields and displacements indicate the direction of changes to be made in the geometry of components, use of other materials and connection types. The results of fatigue tests of ABS material and an extended description of the entire MES model of the exoskeleton will be the subject of a separate publication.

Acknowledgement. Work done in the scope of the grant OPUS 9, No. 2015/17/B/ST8/01700

References

1. Kato, Y., Chavez, J., Yamada, S., Hattori, S., Takazawa, S., Ohuchi, H.: A large knee osteochondral lesion treated using a combination of osteochondral autograft transfer and second-generation autologous chondrocyte implantation: a case report. Regen. Ther. **10**, 10–16 (2019)
2. Hol, A.M., van Grinsven, S., Lucas, C., et al.: Partial versus unrestricted weight bearing after an uncemented femoral stem in total hip arthroplasty: recommendation of a concise rehabilitation protocol from a systematic review of the literature. Arch. Orthop. Trauma Surg. **130**, 547–555 (2010)
3. Chen, B., Zi, B., Wang, Z., Qin, L., Liao, W.: Knee exoskeletons for gait rehabilitation and human performance augmentation: a state-of-the-art. Mech. Mach. Theory **134**, 499–511 (2019)
4. Yatsun, A., Jatsun, S.: Modeling quasi-static gait of a person wearing lower limb exoskeleton. In: Radionov, A. (eds.) Proceedings of the 4th International Conference on Industrial Engineering. Lecture Notes in Mechanical Engineering, pp. 565–575. Springer, Switzerland AG (2019)
5. Kong, K., Bae, J., Tomizuka, M.: A compact rotary series elastic actuator for human assistive systems. IEEE/ASME Trans. Mechatron. **17**(2), 288–297 (2011)
6. Grimmer, M., Quinlivan, B., Lee, S., Malcolm, P., Rossi, D., Siviy, C.J., Walsh, C.: Comparison of the human-exosuit interaction using ankle moment and ankle positive power inspired walking assistance. J. Biomech. **83**(23), 76–84 (2019)
7. Guncan, B., Unal, R.: ANT-M: design of passive lower-limb exoskeleton for weight-bearing assistance in industry. In: Carrozza, M. et al. (eds.). International Symposium on Wearable Robotics. Wearable Robotics: Challenges and Trends, Pisa, Italy, pp. 500–504 (2018)
8. GuiFang, Y., Jigui, Z., FuDe, W., Rui, H., Fei, L.: Technical analysis of auxiliary exoskeleton robot based on wheeled platform. In: Proceedings of the Chinese Automation Congress CAC'2017, Jinan, China, pp. 7560–7564 (2017)
9. ADINA: Theory and Modeling Guide, vol. 1. ADINA R&D, Inc., Watertown (2007)

10. Bathe, K.J.: Finite Element Procedures. Prentice-Hall, Inc. Simon & Schuster/A Viacom Company, Upper Saddle River (1996)
11. Śpiewak, S.: Methodology for calculating the complete static carrying capacity of twin slewing bearing. Mech. Mach. Theory **101**, 181–194 (2016)
12. Johson, K.L.: Contact Machanics. Press Syndicate of the University of Cambridge, Cambridge (2004)
13. SKF Group: Rolling bearings (2013)
14. Zhang, H., Cai, L., Golub, M., Zhang, Y., Yang, X., Schlarman, K., Zhang, J.: Tensile, creep, and fatigue behaviors of 3D-printed acrylonitrile butadiene styrene. J. Mater. Eng. Perform. **27**(1), 57–62 (2018)

Modelling of Concurrent Development of Assembly Process and System

Jan Duda(✉)

Institute of Production Engineering,
Cracow University of Technology, Cracow, Poland
duda@mech.pk.edu.pl

Abstract. Industry 4.0 is a way of "designing – producing – delivering – servicing – utilising" products in order to satisfy individualized customer needs and meeting company expectations to "make profit". Basic characteristic components of this revolution paradigm for production are rapid product development in Cyber Physical System and collaborative networks using modern development strategies Concurrent Engineering and Cross Enterprise Engineering.

For rapid manufacturing, flexible and reconfigurable production and integrated approach to the design of production processes and systems is necessary. The paper presents a systematic approach to formal description of the system and assembly process. In the paper methodology of concurrent development and formal description of products, processes and assembly systems using PLM tools was presented. Integration of the technical production preparation phases covers various planning actions, including the ones executed with the computer systems like Design for Assembly (DFA) analysis, MTM time analysis, and line balancing.

Keywords: Concurrent Engineering · Process and assembly system · Computer integration · PLM

1 Introduction

According to the new development strategies, the product development focuses on [1, 4, 5]: as much as possible parallel execution of all development related product life cycle phases, thus creating CE (Concurrent Engineering) environment. CE strategy assumes the development of resources and production facilities at the early product design phases to shorten the production start-up time. Effectiveness of the CE strategies results from the knowledge and better management of information in production process. The efficiency of activities depends on the provision of right information, to right places, right peoples and in the right time. The integrated systems must fulfil the market requirements for the new products and client requirements in relatively short time. The nowadays trends in the computer aided design and management include:

– The methods for the integration of the formal usage of knowledge in the computer aided design and management [2, 8, 18, 20]. This has led to the development of intelligent CAD/CAM systems.

M. Majewski and W. Kacalak (Eds.): IIRTS 2019, LNME, pp. 118–129, 2020.
https://doi.org/10.1007/978-3-030-37566-9_11

- The organization of the planning process including the cooperation of different experts performing the engineering and business processes in the geographically distributed environment. This is related to the informational integration, important from the point of view of products, process and production system development strategies, which necessitates the improvement of the processes of information management during the production.
- The development of integrated product development systems evolves to Product Lifecycle Management (PLM) solutions. PLM environment includes the following applications for the product development: PPM (*Product and Portfolio Management*), CAD/CAM *(Computer Aided Design/Computer Aided Manufacturing)*, MPM *(Manufacturing Process Management)*, PDM (*Product Data Management*). MPM systems have an integrating role in the area of construction, technology and organization of production preparation. The MPM system creates a set of applications that operate on relational data repositories about products, processes and resources - PPR (Product Process Resources) which, by referring to stored data, enable analyses and reports that are the basis for improvement of the production system [9, 11, 14–16].

2 System Approach for Production

In a general systematic approach, the following functional subsystems, production units *JP*, can be distinguished in the production system: preparation of production, material supply, calculation and production planning. They consist of lower subsystems JP_{R-1}. For example, the manufacturing subsystem consists of processing, assembly subsystems, etc. The organizational design is represented by *SP* structure of the production system:

$$SP = \{JP_R, \{JP_{R-1}, \ldots \{JP_{R-r} \ldots\}, \ldots \{JP_0, \ldots\}, \ldots\}, \ldots\} \tag{1}$$

where:

r – level of decomposition of the production system,
JP_{R-r} – production units on r - decomposition level,
JP_0 – production stands.

The production structure *SP* consists of production units with different degrees of complexity. Production unit *JP* is a type of organizational unit; a team of people equipped with work resources, capable of acting independently to implement a specific program. At a specific moment t, there are cooperative, transport, information links, etc. between production units. The production structure is also changing over the time as a result of adaptation processes to the changing external environment and dynamically changing conditions stimulating the development directions of the enterprise. Thus, the production system can be described using the four elements:

$$SPR = (SP, WP, \gamma, t) \tag{2}$$

where:

SP – production structure,
WP – a set of links between the elements resulting from the relations of products, resources and production processes,
γ – mapping on the set of production units,
t – moment in time.

$$\gamma : JP \times JP \rightarrow WP \tag{3}$$

Decomposition of the structure of the production system is described in the paper [8]. In the R-level production unit of the manufacturing system JP_R^M, the production units on the R-1 level can be separated: processing systems implementing manufacturing processes for components of products JP_{R-1}^O, and assembly systems that perform assembly processes of components JP_{R-1}^M. The production unit on the lowest level are the elementary production station JP_0. Further considerations will concern assembly subsystems and processes that take place in them. Due to the variety of products $\{WR\} = \{WR_1, WR_2 \ldots\}$ manufactured in the assembly subsystem $JP_{R-1}^M\{WR\}^M$ of production system SPR, there may be various types of production organization and organizational forms defining the varieties of production organization production line and cell (Fig. 1).

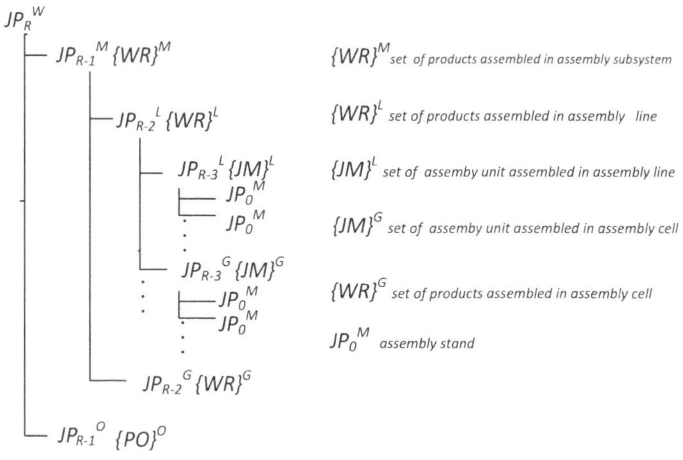

Fig. 1. Hierarchical structure of the manufacturing system JP_R^M.

The assembly process is carried out in the production unit JP^{M}_{R-1}. Assembly process *PTM* for the product is described by:

$$PTM = \{E_{TRM}, E_{OPM}, E_{IDM}\}, SPM \tag{4}$$

where:

E_{TRM} – assembly process actions causing permanent connection of parts,
E_{OPM} – assembly process actions changing the location of the object in the system of assembly tools, assembly units, assembly equipment,
E_{IDM} – control actions checking the quality of the executed assembly tasks in relation to the requirements in the process documentation,
SPM – structure of assembly process plan.

The actions $E_{TRM}, E_{OPM}, E_{IDM}$ can be described by three blocks of information: action model, execution methods and parameters. The structure of assembly process plan *SPM* has characteristic structure elements and their order depending on:

- features of product, its dimensions, and dimension constraints of location of assembly units,
- requirement no to use in the assembly operation the parts already mounted,
- size of manufacturing batch.

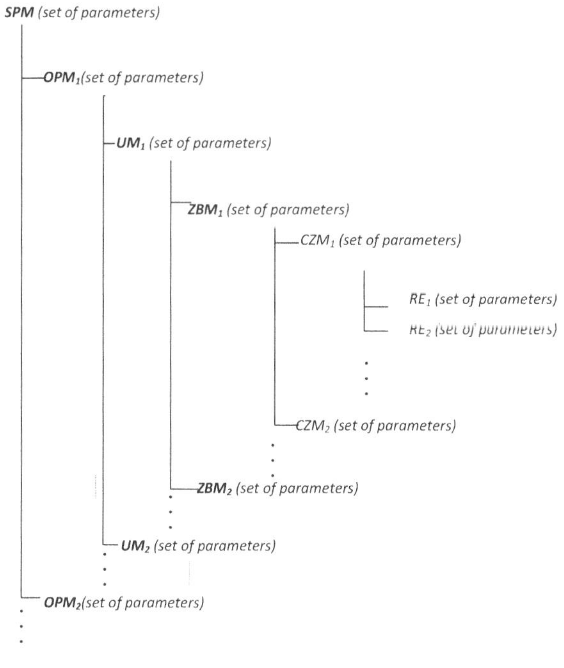

Fig. 2. Structure of assembly process plan. ***SPM*** = [***OPM*** - assembly operation [***UM*** - set-up [***ZBM*** - assembly treatment- [***CZM*** - assembly action- [***RE*** - elementary motion]]]].

Assembly process plan is the result of the planning procedure, which determines the manufacturing actions and the structure of assembly process *SPM*, allowing to receive the design form of the product proposed by the designer.

3 Description of Integrated Process and Assembly Planning

Features of modern development strategies indicate the need for product development phase integration. Integration and parallel execution of activities were received through the separation of the conceptual design stages, allowing for the creation of the variant solutions. Variants are then evaluated in the view of the requirements of the next development phase [7, 8]. The integrations of process and assembly system development suggests the separation of the following phases:

- conceptual process planning,
- detailed process planning,

and

- conceptual production organization planning,
- detailed production organization planning.

The execution of the product development during the production preparation using CE aids the PLM application for the product development management. The concurrent product development is executed by the design, process planning and organization planning teams. The work [10] describes the sub-processes for the conceptual and detailed design and manufacturing process planning using the CAx and DFx methods and systems in BPMN notation. Integration of the design, process planning, and production organization phases was received through the use of computer aided systems (Fig. 3).

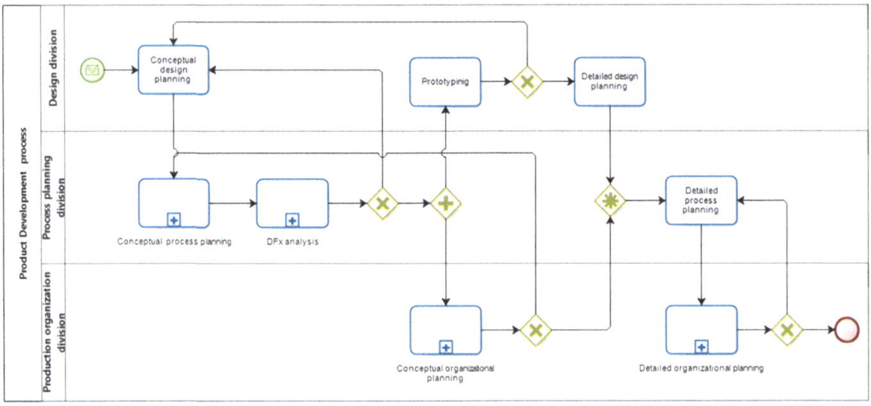

Fig. 3. General production preparation process in the BPMN notations created with Bizagi process modeler software [10].

3.1 Conceptual Process Planning

The assembly process is created on the basis of the digital product model prepared with CAD system. The process planning activities include:

- development of the product assembly structure – separation of the assembly units (assemblies, subassemblies and parts) Fig. 4 step 1,
- development of the assembly process plan including the basic parts for separated assembly units, methods and hierarchical order of assembly of these units to receive the design features of the product,
- definition of the assembly tasks and their characteristics Fig. 4 step 3.

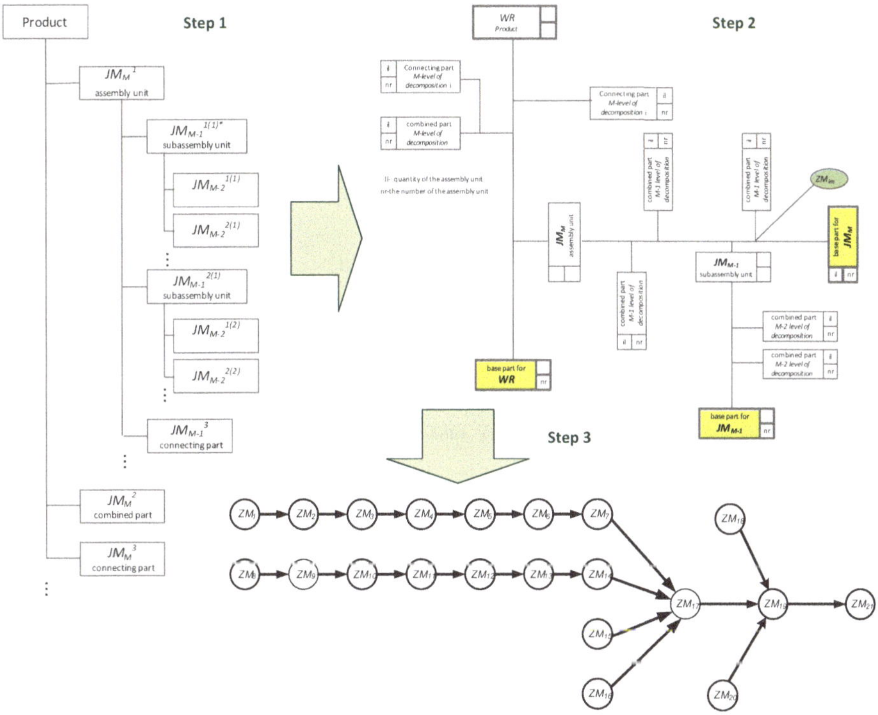

Fig. 4. Steps of conceptual process planning.

The main assembly tasks (basic and special) ZM influence the type and degree of functional properties of the assembly units and constitute the essential content of each process:

$$ZM = \{ZM_1, ZM_2 \ldots ZM_{izm} \ldots ZM_{lzm}\} \tag{5}$$

where:

lzm – the number of assembly tasks in the assembly process.

The main assembly task is defined by the type and method of connection implementation and includes: mutual orientation, connection (mating - folding), fix connection [17]. Special tasks cause the unit to acquire higher-order functional properties. They arise after making a particular connection influencing it (for example tightening, balancing). The main assembly tasks are complemented by auxiliary tasks, which are divided into:

- preparatory tasks, performed before assembly to prepare parts or assemblies for connections,
- ending tasks, occurring after assembly, which aim is to give the final appearance, protection against, for example, corrosion and preparation for shipment.

The set of assembly tasks is supplemented by control tasks separating the incorrectly assembled units, blocking the units not properly assembled to be transported to the next assembly stage or the product. They are divided into preliminary, inter-operative and final. On the set of assembly tasks, the graph *GOK* can be describing (Fig. 1 step 3)

$$GOK = \langle ZM, RK, \mu \rangle \tag{6}$$

where:

ZM – set of assembly tasks,
RK – set of precedence relations,
μ – mapping on the set of assembly tasks.

$$\mu : ZM \times ZM \rightarrow RK \tag{7}$$

Precedence relation can be described by the matrix $RK\left[rk_{ij}\right]_{n \times n} i,\ j = 1..n$

$$RK\left[rk_{ij}\right]_{n \times n} i,\ j = 1..n \tag{8}$$

$$rk = \begin{cases} 1\, when\, task\, i\, predes\, task\, j \\ 0\, in\, the\, other\, case \end{cases}$$

The multivariant nature of the assembly processes is due to the possibility to use several assembly sequences with the application of methods and manufacturing means and different automation levels. On the base of the defined variants of manufacturing/assembly processes, the manufacturability analysis DFM and assemble-ability analysis DFA are carried out [5, 6]. The results of the subsequent iterations are used to simplify the product design (by minimization of the number of parts and the integration of parts), thus decreasing the time and costs of the assembly and to estimate the times and costs for different manufacturing methods of constituent product elements. DFA classify the degree of difficulty of assembly actions and uses it for the determination of the assembly time of all assembly task T_{izm}. The time determined by DFA [12] methodology results from the classification of design features influencing the execution of assembly actions. On this planning phase, it is possible to evaluate the total product assembly time TM.

$$TM = \sum\nolimits_{izm=1}^{lzm} T_{izm} \tag{9}$$

The product design resulting from the subsequent iterations and its assembly plan form the base for defining the graph of the assembly activities, representing the admissible variants of the execution of product assembly.

3.2 Conceptual Production Organization Planning Stage

The conceptual organization stage is used to select the appropriate form of the production organization, production pace, and for the initial calculation of the number and type of functional subsystems. On this stage, also the type and organizational form of assembly system are selected. The main goal of the conceptual organization planning is the development of the manufacturing system concept for planned for assembly of products. Decision taken on this stage can be changed later and should be detailed on the subsequent phases. Nevertheless, they influence further solutions. The base for taking these decisions is limited information gathered during the previous development activities, describing:

- assortment specified by the set of items $\{WR\} = \{WR_1, WR_2 \ldots\}$,
- production program for each assortment item,
- developed in the conceptual phase of technological design, the concepts of technological processes of assembling assortment positions.

In the first design step, the production structure of the assembly system $JP_{R-1}^{M}\{WR\}^{M}$ (Fig. 1) is determined. The established number and type of functional subsystems is the basis for conducting organizational arrangements within the isolated subsystems. Further steps are dependent on the adopted variation of the production organization in the dedicated production assembly subsystem. The procedure leading to the design of the assembly line will be presented below. In terms of conceptual organizational design, the number of assembly stations for the planned production volume is determined. Input data for the calculation of the production cycle are required production volume Pr and acceptable production time Tp:

$$Ts = \frac{Tp}{Pr} \tag{10}$$

The conceptual phase of the assembly process delivers:

- information on all tasks that must be performed as part of the process ZM (Eq. 13),
- times of implementation of assembly tasks T_{izm},
- graph of GOK sequence restrictions (Eq. 6),
- total time of performing all TM tasks (Eq. 9).

Based on the above data, the theoretical minimum number of assembly positions from the dependencies can be calculated:

$$LS\mathrm{min} = \frac{\sum_{izm=1}^{lzm} T_{izm}}{Ts} \tag{11}$$

where:

T_{izm} – time needed to complete the i-th task,
lzm – the number of assembly tasks.

Taking into account the sequence relation specified by the *GOK* graph imposed on the assembly tasks, it is possible to pre-allocate assembly tasks to stands. For this purpose, all *ZM* assembly tasks should be grouped into such subsets, so that they create assembly operations carried out on the assembly line workplaces. It is required to meet the postulate of completeness of assembly:

$$\bigcup_{im=1}^{lm} ZM_{im} = ZM \tag{12}$$

where:

lm – the number of assembly stands (operations).

It is also important that no assembly task belongs to two different subsets, as the same assembly task cannot be performed on two positions. If the labour consumption of assembly tasks is higher than the average cycle, several workstations should be placed parallel to synchronize the assembly line. The number of these workplaces is obtained by dividing the labour-intensity of individual operations by the average value of the line measure and rounding the result to full units:

$$Ls_m = \frac{T_{im}}{Ts} \tag{13}$$

where:

Ls_m – the number of stands planned for the implementation of the operation.

The outcome of these actions is the conceptual assembly system forming the base for the subsequent detailed process planning. Next project steps, implemented in the phase of conceptual organization planning stage, capture a number of design activities leading to the development of technical solutions for both elementary production units JP_0 forming the system as well as solutions in the field of interstate transport within the designed subsystems [3, 13]. The result of the project activities is layout justified by technological reasons (deployment and design of assembly stands) for which the structure of the assembly process will be designed.

3.3 Detailed Process Planning Stage

The implemented project activities include:

- determination of the structure of assembly operations, cycles and activities,
- development of control programs for assembly devices, manipulators,
- standardization to determine the technical time standard for assembly operations,
- development of instruction sheets for assembly operations.

For the set of $\{E_{TRM}\}$ activities planned to be carried out at the designed assembly station determined as a result of technological design, the technological parameters of the assembly connection process are determined, based on which the selection or design of equipment and assembly tools is carried out. Requirements are established for the orientation and mounting of the *JM* assembly units required for the proper selection of the assembly tooling. In this way, the set up *UM* are set in the structure of the operation, and *ZBM* assembly treatment are included in the set-up (Fig. 2). The assembly treatment is written on the main action *CZM*, which results in the connection of units and auxiliary actions, e.g. loading and orientation of the assembly units [19]. The technological parameters of the assembly connection are determined for the main actions. Actions are written out for elementary movements. In the case of operations performed by numerically controlled assembly and supply devices, project activities include the development of control programs for these devices. For separated structural elements of operations, main and auxiliary times are calculated using analytical and computational methods on the basis of technological standards and norms of elementary movements. The result of the design is the assembly process *PTM*, described by the set of assembly activities $\{E_M\}$; $\mathrm{E_{TRM}}$ transformers, E_{OPM} operators and identifiers E_{IDM} realized by assembly system elements, as a result of which parts are joined into higher order assembly units (Eq. 4). The creative decision process is also the structure of the assembly process technological process defined by the assembly process and attributes of the assembly process characterized by:

$$SPM = OPM_1(SM_1, tpz_1, tj_1, \ldots), OPM_2(SM_2, tpz_2, tj_2, \ldots)\ldots \quad (14)$$

included in the manufacturing documentation in the form of a manufacturing assembly sheet. For each operation, instruction and standardization sheets are created with the characteristics of set-up, treatments, action and elementary movements with the required degree of detail, depending on the production conditions (for example the type of the production organization). The technological processes of *PTM* assembly established for assortment (products) form the basis for organizational improvement of the positions forming the assembly subsystem and the implementation of the detailed organizational design phase.

3.4 Detailed Production Organization Planning Stage

In the implementation of the detailed phase of the organizational design of the assembly system, factors describing its configuration and functional status, deciding about success or failure during the exploitation of the designed system should be taken into account.

The group of factors constituting the input data in the detailed organizational design stage are the attributes concerning the allocation and configuration of resources in separate production units. Developed assembly process plans *PTM* of products planned for assembly in the designed system are the basis for choosing the organizational variant of the assembly subsystem and the basis for the implementation of the detailed organizational design phase.

The chosen structure of the assembly process and times of assembly tasks forming operations are synchronized in the designed assembly subsystem, which depends on the adopted form of production organization. On the production lines, these steps include recalculating the number of positions and balancing lines.

4 Conclusion

The proposed methodology of concurrent development and formal description of products, processes and manufacturing systems using such PLM tools increases the integration of the technical production preparation phases and covers various planning actions, including the ones executed with the computer systems like Design for Assembly (DFA) analysis, MTM time analysis, and line balancing.

Nevertheless, it should be noted the PLM solutions are only the tools speeding up the organizational and process planning stages of production preparation. The key condition to receive the good results are high qualifications of designer, process planners and manufacturing organization engineers developing the system components. Additional research works are required especially for the conceptual and detailed production organization planning phases. Suggestions for the order of design actions depending on the characteristics of the process and manufacturing system shall be elaborated. Especially promising is the use of the Artificial Intelligence (AI) techniques for the initial estimation of the manufacturing concepts. Nevertheless, despite the huge number of advanced computer systems for various production preparation phases, the most important decisions are still taken by experts.

References

1. Adam, O., Hoffer, A., Zang, S., Hammer, C., Jerrentrup, M., Leinenbach, S.: A collaboration framework for cross-enterprise business process management. In: Proceedings of International Conference on Interoperability of Enterprise, Software and Applications, INTEROP-ESA 2005, Geneva, Switzerland (2005)
2. Chlebus, E.: CAx Computer Technology in Production Engineering. WNT, Warsaw (2000)
3. Ciszak, O., Żurek, J.: Modelling and simulation of assembly sequence of machine parts and sets. Adv. Manuf. Sci. Technol. **26**(3), 55–62 (2002). PAN, Rzeszów

4. Contreras, J.D., Garcia, J.I., Pastrana, J.D.: Developing of Industry 4.0 applications. Int. J. Online Eng. **13**(10), 30–47 (2017)
5. Duda, J.: Modelling of concurrent development of the products, processes and manufacturing systems in the product lifecycle context. In: New Trends in Technologies: Devices, Computer, Communication and Industrial Systems. SCIYO, Vienna (2010)
6. Demoly, F., Xiu-Tian, Y., Eynard, B., Rivest, L., Gomes, S.: An assembly-oriented design framework for product structure engineering and assembly sequence planning. Robot. Comput. Integr. Manuf. **27**(1), 33–46 (2011). Manuscript Number: RCIM-D-09-00078R1
7. Duda, J.: Computer integrated development of processes and assembly systems. In: Lisowski E., Filo G., Filipowska R., Fabiś-Domagała J. (eds.) Computer Aided Mechanical Engineering, vol. 1, pp. 99–106. Cracow University of Technology Press, Cracow (2011). (Technical Transaction: no. 4-Mechanics)
8. Duda, J.: Management of product development in systematic approach. Cracow University of Technology Press, Cracow (2016)
9. Duda, J.: Formal description of integrated process and assembly system planning. In: Hamrol, A., Ciszak, O., Legutko, S., Jurczyk, M. (eds.) Lecture Notes in Mechanical Engineering, Advances in Manufacturing, Poznan, pp. 79–89 (2017)
10. Duda, J.: Modelling and implementation of product development strategy. Concurr. Eng.: Res. Appl. **26**(21), 187–197 (2018)
11. De Lit, P., Delchambre, A.: Integrated Design of a Product Family and its Assembly System. Springer, New York (2003)
12. Duda, J., Karpiuk, M., Gawąd C.: Design for assembly analysis in SolidWorks system. Engineering design and construction, no. 3 (2010)
13. Grzechca, W.: Final results of assembly line. In: Grzechca W., Assembly Line - Theory and Practice, pp. 3–12. InTech, Rijeka (2011)
14. Eigner, M.: Product lifecycle management – the backbone for engineering. In: Proceedings of the International Conference "Virtual Design and Automation", Poznan (2004)
15. Eversheim, W., Rozenfeld, H., Bochtler, W., Graessler, R.: A methodology for integrated design and process planning on a concurrent engineering reference model. Ann. CIRP **44**(1), 403–406 (1995)
16. Kahlert, T.: From PDM to PLM – from a workgroup tool to on enterprise – wide strategy. In: Proceedings of the International Conference "Virtual Design and Automation". The Publishing House of Poznan University of Technology, Poznan (2004)
17. Łebkowski, P.: Computer-Aided Mechanical Assembly Methods in Flexible Production Systems. AGH University of Science and Technology Press, Cracow (2000)
18. Pokojski, W.J., Shuishi, F., Salwiński, J., New world situation: new direction in concurrent engineering. In: Pokojski, J., Fukuda, S., Salwiński, J. (eds.) Proceedings of the 17th ISPE International Conference on Concurrent Engineering. Springer, London (2010)
19. Puff, T., Sołtys, W.: Fundamentals of Machine and Device Assembly Technology. WNT, Warsaw (1980)
20. Torres, V.H., Rios, J., Vizan, A., Perez, J.M.: Integration of design tools and knowledge capture into a CAD system: a case study. Concurr. Eng. Res. Appl. **18**(4), 311–324 (2010)

A Method for Applying Antipatterns and Neural Networks to Automate Detection of Errors in Designs of Mechanical Constructions

Andrzej Tuchołka(✉), Maciej Majewski, Wojciech Kacalak, and Zbigniew Budniak

Faculty of Mechanical Engineering, Koszalin University of Technology, Raclawicka 15-17, 75-620 Koszalin, Poland
andrzej.tucholka@gmail.com

Abstract. Proposed method allows for early detection of mistakes in designs of mechanical constructions. It is based on a numerical classification of a symbolic representation of construction's features against a set of defined antipatterns (known, incorrect, repeatable data patterns). We present an approach to identify antipatterns described using a symbolic language KXML and a method of intelligent quality assessment enabling calculation of the similarity of the tested element with the antipattern data set. Additionally, we highlight selected properties of numerical models directly supporting analysis of the structure of mechanical constructions.

Keywords: Antipatterns · Quality assessment · Design of machine elements

1 Introduction

Since many years, numerical models (mathematical, physical, algorithmic) are being developed to allow machine designers to better predict the behavior of the construction they are building [8]. Along with the development of the internet and communication technologies, more advanced complex data and processing models are being created that are being widely implemented in mechanical engineering [1–4, 11]. Such models enable classification of machine designs represented with CAD models [16], and many other uses based on the pattern recognition in symbolic data [12–15]. All of these models require modifications such that they can be effectively applied in the machine design.

While describing the capabilities of the classification (similarity grading, distance) of the features of mechanical constructions, it is important to highlight the unique capabilities of pattern recognition, found in new neural network designs (ConvNet, CapsNet) [15, 20–22]. The benefits of these networks arise from novel solutions such as a flexible convolution kernel (window) and capsule routing. Comparing these models with earlier designs like Hopfield or Hamming networks [6], the advantage of the ability to decompose and process the data structure is evident. This also enables the network to detect multiple patterns inside one data packet [15, 20].

M. Majewski and W. Kacalak (Eds.): IIRTS 2019, LNME, pp. 130–138, 2020.
https://doi.org/10.1007/978-3-030-37566-9_12

Other, similar numerical models, allow to automatically group (cluster) the data based on the similarity between distinct features [5, 7, 12, 14, 15]. We were able to apply them to analyze similarity between machine parts, but we did not find any way to include data's structure in the calculations.

2 Identification of Antipatterns

Both, presented numerical models and simulations, attempt to predict behavior driven qualities of the construction. Such models require additional interpretation of their outputs to enable defining the subjective quality of the element.

Referencing antipatterns in the calculations performed by presented numerical models, creates a negative quality reference allowing us to automate calculating a reversed quality grade. Such grade enables reasoning with regards to the quality of the construction.

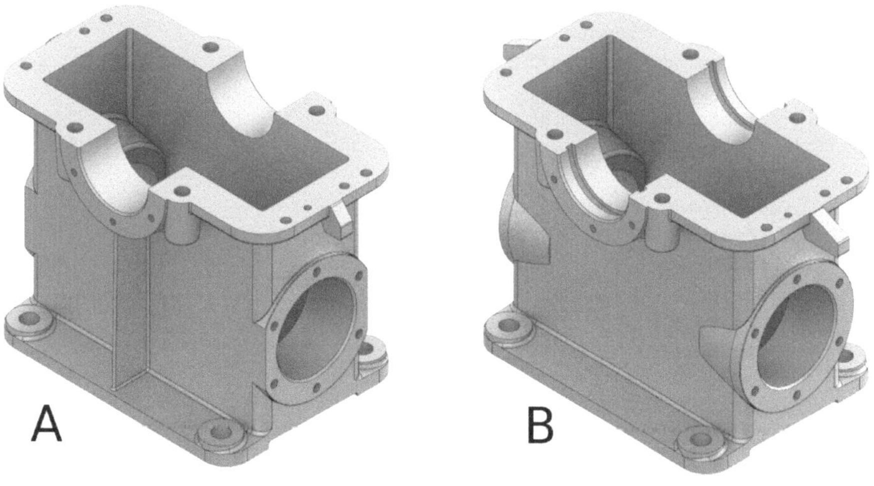

Fig. 1. Examples of two gearboxes: correct (A) and antipattern (B).

Roots of antipatterns can be found in computer science [9], and represent a set of multi-disciplinary knowledge defining repeatable structural mistakes. In the context of mechanical constructions, we identify antipatterns (Fig. 1) in feature values (e.g. incorrect roughness or thickness of a surface), structural relations (e.g. lack of ribbing on the corpse), geometrical (e.g. incorrect limiting outline or perpendicularity of axis).

The quality grading of the machine elements based on the variation from the set of antipatterns [18, 19, 21, 22], can be used both to avoid repeatable design errors, and to automatically compare construction elements based on their similarity to incorrect ones. As long as this method allows to detect design errors that were known earlier, its results should not be treated as exhaustive.

3 Symbolic Description of Construction Features KXML

The challenges arising from normalization of data describing features of a mechanical construction arise mainly from the nature of technical drawings as they have been designed to use and interpret by a trained human. The difficulties also come from variations in data communication styles and a series of visual irregularities: rotations, visibility, positioning, and data description formats.

Another set of challenges is being addressed by a set of existing data formats, allowing to describe constructions. Standards such as ISO 10303 (STEP), ECMA-363, and DXF provide a set of structures that has been designed for visualization and production purposes.

```
<struct class="shaft" id="s001">
  <model class="mechanical">
    <feature id="diameter" unit="mm"/>
    <feature id="length" unit="mm"/>
    <feature id="angle" unit="deg" vector="(0,0,1)"/>
    <node class="mandrel" id="p001">
      <feature id="length">48</feature>
      <feature id="diameter">35</feature>
      <node class="mill">
        <feature id="length">12</feature>
        <feature id="angle">45</feature>
      </node>
      <node class="thread">
        <feature id="length">24</feature>
      </node>
      <node class="undercut">
        <feature id="radius">8</feature>
        <feature id="diameter">8</feature>
      </node>
    </node>
  </model>
</struct>
```

List. 1. KXML representation of a shaft, presenting a general structure of the language.

Within the scope of the presented method, we propose using a symbolic language KXML (List. 1) [17] that builds a description of the construction using a decomposition of nodes, feature definitions and structure, and a lexical classification structure with inheritance. A symbolic object oriented KXML language, allows us to perform the structural decomposition of the construction in a manner simplifying the normalization of data for further calculations. We present the construction's description in KXML as an input, but it is possible to use different notations or languages to present similar data.

We designed KXML aiming to get a synthetic and simple representation of mechanical constructions. The main focus driving the design was placed on the function of an element. The concept element's correctness is external to its description, so when using construction's data, we need to clearly separate the correct designs from antipatterns. Creating a description of a construction starts with its function based decomposition into base elements, to which we later assign attributes with concrete feature values. Such structured description can be compared to another one, but not in a trivial manner due to variations in the constructions and their parameters [19]. Such approach makes it easier to compare elements of mechanical constructions through similarities in their structures and feature values. The lexical class descriptions, enable mapping and comparing features of related design elements (via their function) [10].

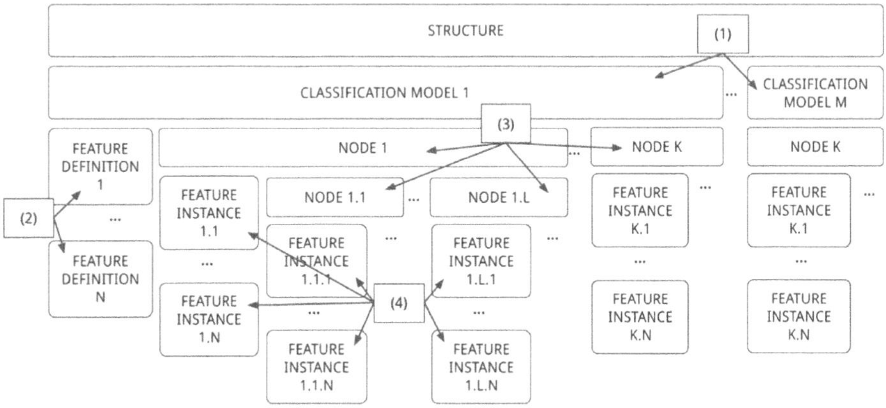

Fig. 2. A general structure of the symbolic description we use to describe constructions.

To describe the construction with KXML we use four main elements (Fig. 2): (1) the classification model – providing us real-world context for the analysis and interpretation of features and structural nodes; (2) definition of the feature – defining the fundamental data structure used to describe the construction; (3) instances of nodes and their structure – explicitly representing construction's decomposition based on the element's function; (4) feature instances – that contain values (or math functions resulting with a value) used to define the similarity of the constructions. Feature instances are analyzed in the context defined by a concrete classification model, their surrounding structures and position in the lexical node structure.

4 Intelligent Quality Grading

To enable automatic detection of mechanical design errors through the similarity to antipatterns (Fig. 3) [18, 19, 21, 22], the classification algorithms have to enable concrete definition of the antipattern set. Such data set should contain data samples focusing on demonstrating the incorrect nature of the design.

The multi-dimensional nature of the mechanical design does not allow for direct application of data formatted in KXML format. Each of the algorithms also requires a different input. The simplest of the models we have tested (Additive, Multiplicative,

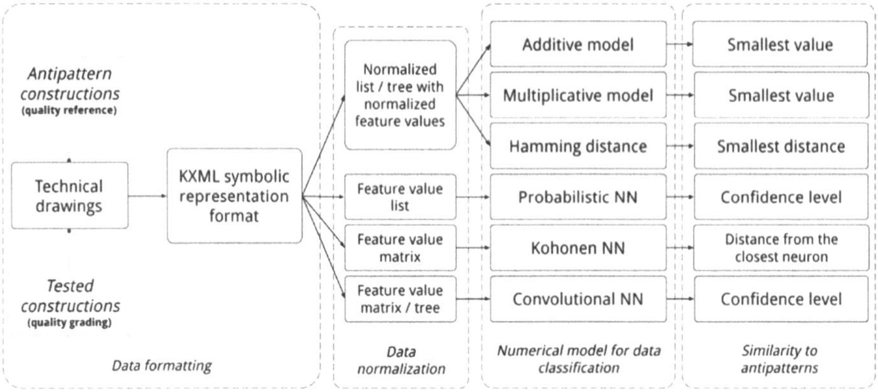

Fig. 3. Analyzed numerical models for fitness in detecting similarity to antipatterns.

Hamming) need to be provided with a one dimensional data structure (list, tree). One their own, these models do not allow to reflect on the structure of the design (with a modest exception of using tree based approximations).

Models using deep neural networks like ConvNet (designed for the purposes of detecting embedded patterns and extraction of structural data in photographs [13, 15, 20]) provide a set of unique features allowing to tune them to the needs of analyzing elements of mechanical designs [18, 21, 23, 24]. These advantages arise from introduction of specific techniques like convolutions, moving kernels, and data filters; enabling dynamic configuration of the network with alignment to the data patterns of the input and training data.

5 Automatic Quality Analysis of Mechanical Constructions

Usage of the KXML symbolic language to represent the construction designs, not only simplifies transitions between multiple data formats required by the models. It also reduces the impact of data variability on the final result, and so allows us to avoid pitfalls of popular CAD/CAM formats. Several attempts have been made [13, 15, 16] demonstrating analysis of the model's structure based off photographs of whole or parts of objects. In fact, these approaches cannot be used in analysis of mechanical constructions due to their focus on the surface of the material. It is probably possible to apply them at detecting surface antipatterns though.

A clear conclusion arising here, is that the ability of the numerical model to attempt at analyzing a mechanical design, depends on the inclusion of the structural information in calculations of the model. To match a model to this need, we have configured a convolutional network design (ConvNet) to allow detection of recurring data patterns in descriptions of mechanical designs. For the results of ConvNet's data classification be useful in error detection, we have to relate its similarity results with quality criteria. For this, we are using a set of known design antipatterns as a training dataset. This approach creates a negative quality reference based on which we can grade the subjective (to the training dataset) quality of a mechanical design.

It is important to highlight the difference between the positive correlation occurring in when classifying against a correct dataset, from the negative one when comparing against antipatterns (Fig. 4). This can lead to further improvements in quality interpretation, by adding weighed grade of similarity to correct designs.

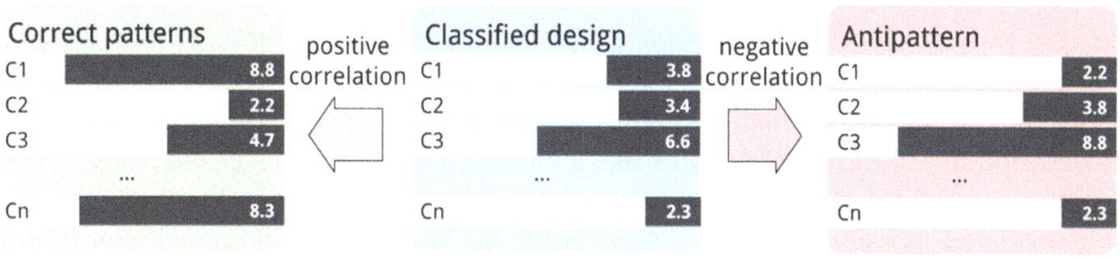

Fig. 4. The positive and negative correlations defining a quality reference.

In the presented method (Fig. 5) the input data contains a matrix of feature values, accompanied by a numerical representation of node class structure. Such model, allows to generate network activation based on local feature values, that through convolutions and sampling layers support identification and classification of patterns found in the symbolic description of the mechanical design.

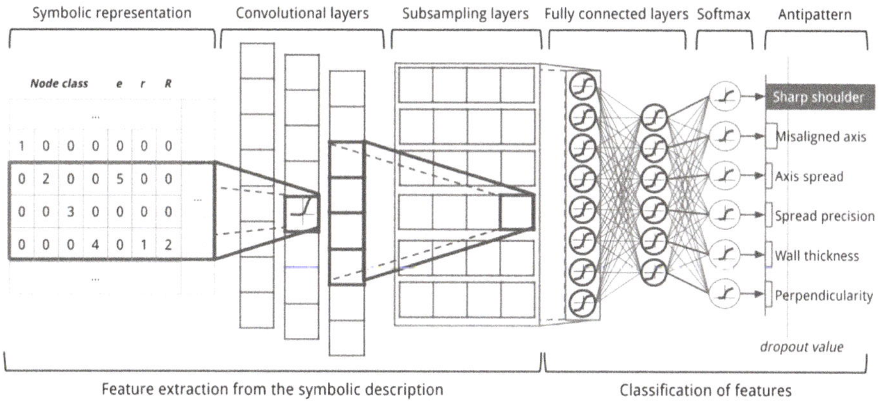

Fig. 5. Model of applying ConvNet for automatic quality analysis.

We are highlighting the ConvNet designs, as among other tested models (Fig. 3), it allows for the highest level of direct association of the performed calculations to the multi-dimensional structure of the symbolic description of the construction. Compared to algebraic models, Hopfield, Hamming, and probabilistic networks, algorithmic extensions introduced by convolutional neural networks increase the depth of analysis in terms of structure and position of feature values in descriptions of mechanical designs [6, 7, 13, 14, 18, 19].

It is worth noting, that there is also a possibility of applying Kohonen maps [5] (or other self organizing feature maps) to classify feature patterns and so detecting mistakes in machine elements. In this case it is necessary to create a map for each of the antipatterns, that would then allow to compare distances between closest points on the antipattern maps and the map of the tested design [25]. This approach seems promising in terms of precision and seductiveness with which the maps can be compared. Still, it does not provide direct features supporting analysis of the structure. Another problem with applying Kohonen maps for automated quality analysis is the amount of calculations that grows exponentially with the number of patterns and comprehensiveness of the description [26]. Lack of features like a moving kernel also makes the analysis very complex and rigid due to the need for predefined feature placement in all of maps, for all considered possibilities. It is foreseeable to implement additional algorithmic (possibly ConvNet and CapsNet inspired) techniques increasing the feasibility of self organizing maps for automatic quality analysis.

6 Conclusions

Calculation models basing on neural networks create a unique opportunity to further automate the quality analysis of mechanical constructions. The main challenge facing the adoption of neural models in mechanical design is data normalization to match the requirements of modern machine learning algorithms. Compared to algebraic or other deterministic models, neural networks require additional configuration to adapt their features and layer structure to the contents and hierarchy of construction's symbolic description.

The benefits of using neural networks to grade the quality of mechanical constructions become visible in convolutional networks, arise from their ability to detect local features in symbolic descriptions of machine element designs. Localized pattern detection allows to reason with regards to the structure of the construction. Multi-layered ConvNet design allows to process relations occurring between data patterns found in the whole scope of the description. The ability to adapt filters used by the network to highlight specific feature combinations, kernel sizes and other configuration, allow for further tweaks increasing the adaptation of the model to analyzed structures. On the other hand, the excessive application of fully connected or sub-sampling layers, decreases the ability of the network to classify complex machine elements [21, 22].

Advances in the field of machine learning [15, 20], pattern recognition [13–16], data dimension reduction [12], created an opportunity to automate processes aiming to eliminate repeating errors (i.e. antipatterns) in mechanical construction designs.

Proposed method of applying antipatterns allows, in a unique way, to create a reference quality layer based on a library of antipatterns. A quality reference is necessary to automate calculations interpreting the fitness of elements with regards to a specific function. A composition of benefits arising from ConvNet design with an antipattern based quality reference, enable automatic interpretation of results by grading their quality through similarity to antipatterns [18, 19, 21, 22].

6.1 Further Research Opportunities

Aside from the potential lying in possible extensions to the self organizing maps, it is worthwhile to analyze the unique model of grouping the neurons in capsules [20]. This approach allows for better representation of concrete elements by allowing for more flexibility in the structure of feature descriptions. Taking into account strengths of the neural models that lie in the ability to detect structural patterns in data, proposed capsule routing mechanism might increase the effectivity of using deep networks in detecting antipatterns in mechanical construction designs.

References

1. Duer, S.: Diagnostic system with an artificial neural network in diagnostics of an analogue technical object. Neural Comput. Appl. **19**(1), 55–60 (2010)
2. Duer, S.: Diagnostic system for the diagnosis of a reparable technical object, with the use of an artificial neural network of RBF type. Neural Comput. Appl. **19**(5), 691–700 (2010)
3. Duer, S., Duer, R.: Diagnostic system with an artificial neural network which determines a diagnostic information for the servicing of a reparable technical object. Neural Comput. Appl. **19**(5), 755–766 (2010)
4. Duer, S.: Artificial neural network in the control process of object's states basis for organization of a servicing system of a technical objects. Neural Comput. Appl. **21**(1), 153–160 (2012)
5. Kohonen, T.: Self-organized formation of topologically correct feature maps. Biol. Cybern. **43**(1), 59–69 (1982)
6. Lippmann, R.P., Gold, B., Malpass, M.L.: A comparison of Hamming and Hopfield neural nets for pattern classification. Massachusetts Institute of Technology, Lincoln Laboratory (1987)
7. Pearl, J.: Probabilistic Reasoning in Intelligent Systems: Networks of Plausible Inference. Morgan Kaufmann Publishers, Burlington (1988)
8. Weld, D.S., de Kleer, J.: Readings in Qualitative Reasoning About Physical Systems. Morgan Kaufmann Publishers, Burlington (1990)
9. Koenig, A.: Patterns and antipatterns. J. Object-Oriented Program. **08**, 46–48 (1995)
10. Lung-Wen, T.: Mechanism Design: Enumeration of kinematic Structures According to Function. CRC Press, Boca Raton (2001)
11. Knosala, R.: Applications of artificial intelligence methods in production engineering. WNT, Warsaw (2002). (in Polish)
12. Vazirani, V.V.: Approximation algorithms. WNT, Warsaw (2005). (in Polish)
13. Bishop, C.M.: Pattern Recognition and Machine Learning. Springer, Heidelberg (2009)
14. Zhou, Y., Cheng, H., Xu Yu, J.: Graph clustering based on structural/attribute similarities. J. Proc. VLDB Endow. **2**(1), 718–729 (2009)
15. Farabet, C., Couprie, C., Najman, L., LeCun, Y.: Learning hierarchical features for scene labeling. IEEE Trans. Pattern Anal. Mach. Intell. **35**(8), 1915–1929 (2013)
16. Fei-wei, Q., Lu-ye, L., Shu-ming, G., Xiao-ling, Y., Xiang, C.: A deep learning approach to the classification of 3D CAD models. J. Zhejiang Univ. Sci. **15**(2), 91–106 (2014)
17. Tuchołka, A., Majewski, M., Kacalak, W.: Object-oriented, symbolic notation for design features, relations and structures. Mach. Eng. **1**(20), 112–120 (2015)

18. Kacalak, W., Majewski, M., Tuchołka, A.: Intelligent assessment of structure correctness using antipatterns. In: The Proceedings of the International Conference on Computational Science and Computational Intelligence CSCI'2015, Las Vegas, pp. 559–564. IEEE Xplore Digital Library (2015)
19. Kacalak, W., Majewski, M., Tuchołka, A.: A method of object-oriented symbolical description and evaluation of machine elements using antipatterns. J. Mach. Eng. **16**(4), 46–69 (2016)
20. Sabour, S., Frost, N., Hinton, G.E.: Dynamic routing between capsules. Computer Vision and Pattern Recognition. arXiv:1710.09829 (2017)
21. Tuchołka, A., Majewski, M., Kacalak, W., Budniak, Z.: A method for intelligent quality assessment of a gearbox using antipatterns and convolutional neural networks. In: Silhavy, R. (ed.) CSOC 2018. Advances in Intelligent Systems and Computing, vol. 764, pp. 57–68. Springer, Cham. (2018)
22. Tuchołka, A., Majewski, M., Kacalak, W., Budniak, Z.: Comparison of numerical models used for automated analysis of mechanical structures. In: Silhavy, R., Silhavy, P., Prokopova, Z. (eds.) CoMeSySo 2018. Advances in Intelligent Systems and Computing, vol. 859, pp. 341–352. Springer, Cham. (2019)
23. Kacalak, W., Majewski, M., Budniak, Z.: Worm gear drives with adjustable backlash. J. Mech. Robot. **8**(1), 014504 (2015)
24. Kacalak, W., Majewski, M., Budniak, Z.: Innovative design of non-backlash worm gear drives. Arch. Civil Mech. Eng. **18**(3), 983–999 (2018)
25. Majewski, M., Kacalak, W.: Smart control of lifting devices using patterns and antipatterns. In: Silhavy, R., Senkerik, R., Kominkova Oplatkova, Z., Prokopova, Z., Silhavy, P. (eds.) CSOC 2017. Advances in Intelligent Systems and Computing, vol. 573, pp. 486–493. Springer, Cham (2017)
26. Kacalak, W., Budniak, Z., Majewski, M.: Computer aided analysis of the mobile crane handling system using computational intelligence methods. In: Silhavy, R., Silhavy, P., Prokopova, Z. (eds.) CoMeSySo 2017. Advances in Intelligent Systems and Computing, vol. 662, pp. 250–261. Springer, Cham (2018)

Application of Multi-valued State Assessment in an Intelligent System Diagnosing Hybrid Power System Devices

Stanisław Duer(✉)

Koszalin University of Technology, Raclawicka 15-17, 75-620 Koszalin, Poland
stanislaw.duer@tu.koszalin.pl

Abstract. The present article covers the idea of the examination of the value of the k-th logics of diagnostic information related to the assessment of the states of complex technical items. For this purpose, an intelligent diagnostic system was presented whose particular property is the possibility to select any k-th logic of inference from set $\{k = 4, 3, 2\}$. An important part of this study is the presentation of theoretical grounds that describe the idea of inference in the multi-valued logic examined. Furthermore, it was demonstrated that the permissible range of the values of the properties of diagnostic signals constitutes the basis of the classification of states in multi-valued logic in the DIAG 2 diagnostic system. For this purpose, a procedure of the classification of states in selected values of multi-valued logic was presented and described. An important element in the functioning of diagnostic systems, i.e. the module of inference was presented, as well. The rules of diagnostic inference were characterized and described based on which the process of inference is realized in the system.

Keywords: Technical diagnostic · Intelligent system · Artificial neural networks · Multi-valued state · Knowledge base · Diagnostics information

1 Introduction

In the case of the organization of maintenance activities, the recognition is important of those states in the object which directly precede the occurrence of a fault state, which is partial usability state. Therefore, the use of two-valued logic for the needs of the organization of the maintenance process is insufficient [1, 7, 11–14]. For this reason, three-valued logic that was developed by J. Łukaszewicz began to be applied more commonly in diagnostics [2–5]. The development of diagnostics in three-valued logic is covered in the studies by the Author. The Author has systematically improved the theory and the diagnosing method of technical objects in three-valued logic. Diagnostic inference that is used there distinguishes the partial usability state "1" apart from the usability state "2" and the fault state "0". It was demonstrated in the literature that an introduction of an additional state allows one to increase the diagnostic information that can be obtained, which in the context of the technical maintenance activities of the object increases the organizational, technical and other possibilities of this process [6, 8–11, 19]. Despite the fact that diagnosing issues in three-valued logic are currently

M. Majewski and W. Kacalak (Eds.): IIRTS 2019, LNME, pp. 139–147, 2020.
https://doi.org/10.1007/978-3-030-37566-9_13

being intensively developed, the issues and rules of diagnostic inference developed do not fully solve the problems of modern technical objects. Their high complexity and responsibility of tasks require one to provide the user with quick and reliable information on the technical state of these objects. An introduction of multi-valued diagnostics with a greater number of states is one of the ways to counteract the problem mentioned above. The use of a multi-state classification increases the possibilities to obtain diagnostic information and, as a consequence, the reliability of the technical state of the object examined. Obtaining a relatively low increase of computational costs during an expansion of the number of distinguishable states is an additional impulse to promote the above-mentioned approach. However, the use of multi-valued diagnostics requires many technical and economic analyzes as well as a standardization of principles and rules for the different classes of technical objects [15–21].

The issue of information research carried out in the article developed by the diagnostic system of intelligent diagnoses expressed in two- and three-valued logics is a new subject. There are no descriptions on the subject in the literature. The basis for testing the informability of used logics in intelligent systems is the formula (4) presented in the article. This dependence developed by the author has not yet been published. In the further development of this research problem, work will be conducted in two directions.

The first direction of research development will be related to the improvement of the diagnostic intelligent system in terms of developing the possibility of developing diagnoses in a four-valued logic. It is already assumed that the four-valued logic contains greater possibilities in the recognition of technical states in the studied object.

The second direction of research will be related to the theoretical and practical study of the informativeness developed by the diagnostic system of intelligent diagnoses expressed in the four-valued logic in relation to the already known (applied in practice) two- and three-valued logic. A novelty in this kind of scientific research is a modern, intelligent diagnostic system that realizes the diagnosis of the state in a technical facility, two-, three- and four-valued logics. The diagnostic information obtained this way is assumed to revolutionize the process of handling complex technical objects towards its automation.

2 Diagnostic Application in Intelligent Systems in *k*-Valuable Logics at ($k = 2, 3$)

Technical diagnosis from its beginning of development has always served something. It can be said that the user of the technical object from the diagnostics, or rather from the diagnosis of the technical object, expected information about the state of the object being used. Knowledge of the object's condition is the basis for the organization of the object's service (renewal) system. Another object sought by the user is knowledge in the given diagnostic information is how to use this information effectively in the process of renewing the technical object.

Therefore, it is desired that the acquired diagnoses as a result of the object's diagnostic process contain as much diagnostic information as possible. Hence the concept of "entropy of diagnostic information". This concept can also be presented in the form of a statement about the informability of the diagnosis. It is possible to formulate a definition defining the informability of technical diagnosis.

Definition 1: Informativeness of technical diagnosis is a quantity that defines "how much knowledge - entropy" contains acquired diagnostic information that is the result of diagnosing a technical object [2, 3]. The classic diagnosis of a technical object in binary logic can be presented in the form of dependence:

$$D^{2VL} = <1, 0> \tag{1}$$

where: D^{2VL} - diagnosis expressed in binary logic: "1" - means the usability state of the technical object, "0" - means the fault state.

Diagnosis of technical objects in binary logic is the basis for the development of other types of diagnoses currently referred to in the article as a diagnosis in k-values logics. The basis for the development of diagnostic information is the description of the diagnostic basis for the D^{2VL} logic - divalent, which is shown in Fig. 1. The analysis of the information presented in Fig. 1 shows that the diagnosis of a technical object consists in recognizing changes in the k-th feature of the j-th diagnostic signal $\{X_{i,j}\}$ in relation to the k-th value of its reference signal.

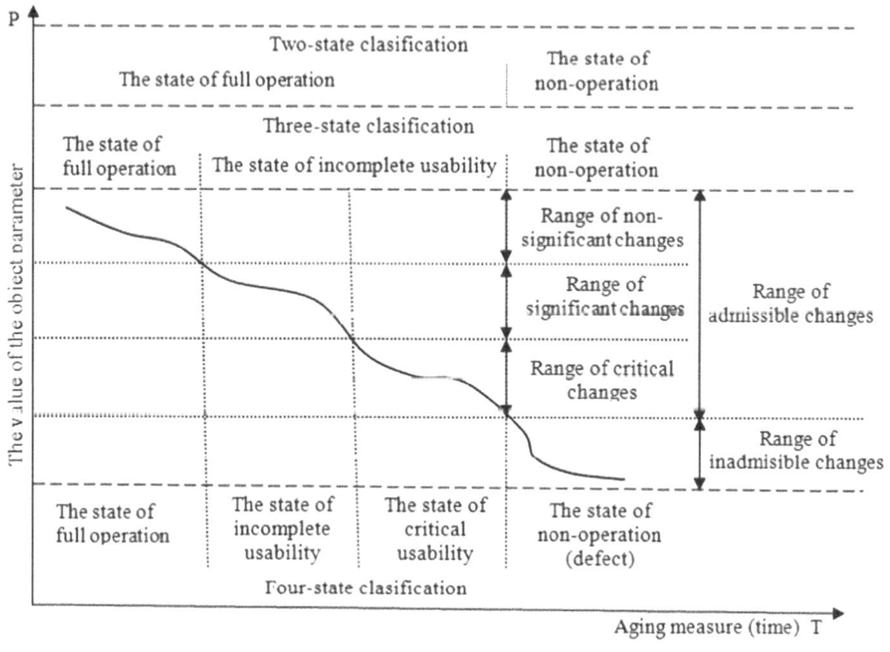

Fig. 1. Diagram of the classification the states of the item in four-valued logic in relation to classification in two- and three-valued logic.

If the changes of this signal characteristic are within the permissible range of possible changes of this k-th feature of diagnostic signal $\{X_{i,\ j}\}$, in Fig. 1 there is a change in the diagnosis in the range from <1> to <0>. Then the application module of the intelligent system will develop the diagnosis "1" - the state of fitness. However, in the case when the changes of this k-th feature of the j-th diagnostic signal go beyond the range of acceptable changes, which interval is called an inadmissible range. For such a situation, the intelligent system application module will develop a diagnosis of "0" - the state of inability [4, 5]. The description of the essence of diagnosing in logic (D^{2VL}) shows that only the range of acceptable changes in the k-th feature of j-th diagnostic signals is the basis for the development of k-valued logic (D^{kVL}). On the basis of diagnostic logic (D^{2VL} - divalent), diagnostics developed using trivalent logic (D^{3VL}). Diagnosis of a technical object in trivalent logic can be presented in the form of dependence:

$$D^{3VL} = <2,\ 1,0> \tag{2}$$

where: D^{3VL} - diagnosis expressed in trivalent logic: "2" – usability state of the technical object, "1" – incomplete usability (partial usability) state, "0" – means fault state.

What was characteristic for the application of three-valued logic was an interpretation of state - "1": the partial usability state.

Definition 2: *Definition of partial usability state* "1": the partial usability state is such a state in the set of the object's states where the object retains the ability to perform its tasks in part. This state occurs directly prior to the occurrence of the fault state.

The issues of diagnosing of technical objects in k-valued logics is still being developed. In the recent years, three-valued logic has been intensively developing in technical diagnostics, especially as regards its applications in intelligent maintenance systems. The Author's studies [2–5] are of significance to the development of three-valued logic (k = 3) where states constitute a set of states $\{2, 1, 0\}$. In the diagnostics of technical objects as well as of industrial and technological processes, three-valued logic has brought about significant information related effects. The determination of the incomplete usability state, i.e. state "1" was a distinctive achievement of three-valued logic. The possibility to recognize the incomplete usability state became grounds for an intensive development of modern intelligent systems that support the organization of the maintenance system of technical objects [4, 5]. It is to be expected that k-valued logics will bring about completely new experiences in the operation of technical objects. Currently, there are no studies in the literature related to k-valued logics for (k > 3) in technical diagnostics.

Based on the analysis of the results of the k-valued logic diagnosis (D^{kVL}), the dependence of diagnosing the informability of diagnoses was developed, which was presented in the form of:

$$E^{VL} = \frac{n^{kVL} - n^{2VL}}{n^{VL}} \cdot 100\% \tag{4}$$

where: E^{VL} - function (entropy) increment of additional diagnostic information, diagnosis expressed in four-valued logic: n^{kVL} - number of states determined in the diagnostic process, for k-value logic, n^{VL} - number of states determined in the diagnostic process for logic (D^{2VL}), n^{VL} - total number of states determined in the diagnostic process.

3 Diagnostics of a Low-Capacity Hybrid Power Station Equipment with Two- and Three-Valued Logic

The functional units of the test object were shown in (DIAG 2) as "units", and the basic components were shown in the same as "elements". The subassemblies are tertiary elements which are "intermediate elements" and enable bidirectional transformation of the hierarchical form of the test object into an internal matrix structure.

The diagnostic status of the test object was determined in (DIAG 2) by testing the set of vectorized diagnostic outputs and comparing them to their respective master vectors [2, 3, 6, 8]. The classification stage during the diagnostic process provided a determination of a defined state of each basic element of the tested engineering object. The classification of the states of the functional units of the test object itself was determined with classification rules and depended on the value of logic (i.e. the number of states). The process of state classification in (DIAG 2) was carried out by opening the "Classification" pane from the tool bar. The "Classification" pane featured configuration and result options, including state tables (maps of states). The configuration allowed setting the type of state logic, the classification interval range, the options of visual output in the state tables, and the RBF artificial neural network parameters. The adopted test object states in multiple-valued logic were determined in the Polish Standards for reliability and servicing quality [15–18] (Figs. 2 and 3).

Figure 4 shows the resultant (and final) form of the diagnostic information with 3-valued logic. The final form of the diagnostic information output for the states of the tested object, the internal structure of which featured 28 basic elements j (modules), located in the functional units i of i. The primary level of the diagnostic information depicted in Fig. 3 for the test object is "Table of basic element states" shown as the values of the states from the set {2, 1, 0}. The secondary and tertiary level of the diagnostic information of the test object could serve to present the states of the functional units i of i, or the state of the test object.

A study of Fig. 4 revealed that the basic elements in the subset {$e_{1.1}$; $e_{1.3}$; $e_{2.1}$; $e_{4.1}$; $e_{5.1}$; $e_{5.2}$; $e_{5.3}$; $e_{6.1}$; $e_{7.1}$; $e_{7.3}$; $e_{7.5}$; $e_{8.1}$; $e_{10.1}$; $e_{12.2}$} were in the operational state "2". The percentage share of the operational elements in the tested object's structure was 50%. The remaining tested subset of basic elements {$e_{1.2}$; $e_{1.4}$; $e_{1.5}$; $e_{2.2}$; $e_{3.1}$; $e_{4.2}$; $e_{5.3}$; $e_{7.2}$; $e_{7.4}$ $e_{7.6}$; $e_{9.1}$; $e_{11.1}$; $e_{12.1}$} were in the partial (reduced) operational state "1". The percentage share of the partially operational elements j of j in the tested object's structure was 46%. Only one basic element {$e_{1.6}$} has the non-operational state "0". The percentage share of the non-operational element in the tested object's structure was 3%.

The analysis of Fig. 4 shows that in the diagnosis of the tested technical object with the 3-value logic for the state of fitness, state - "2", 13 basic elements were classified. The number of elements with a recognized condition of incomplete airworthiness is 14.

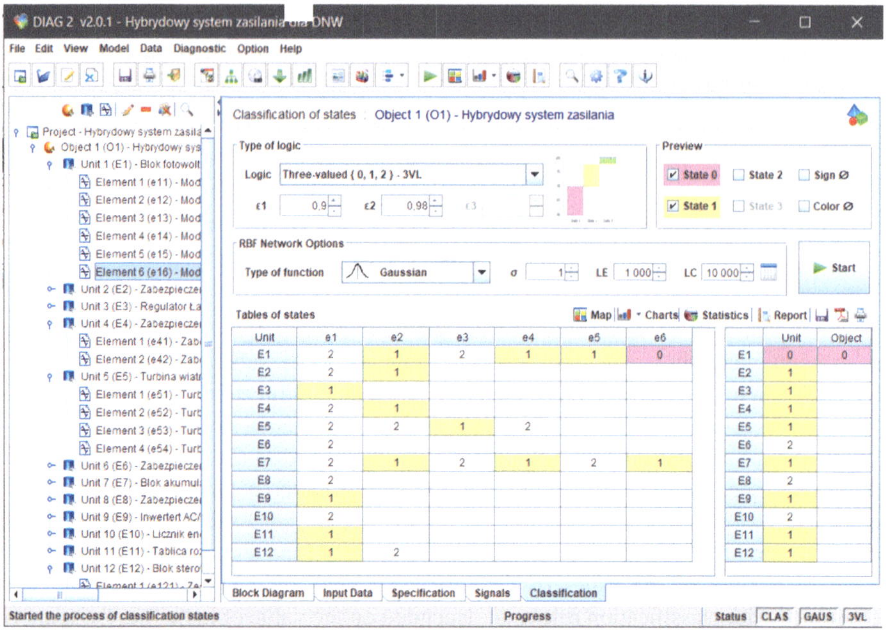

Fig. 2. Resultant form of DIAG 2, “Table of basic element states” in the diagnostic process of the solar power station equipment with 3-valued logic, with: {2} - operational state; {1} - partial (reduced) operational state; {0} - non-operational state.

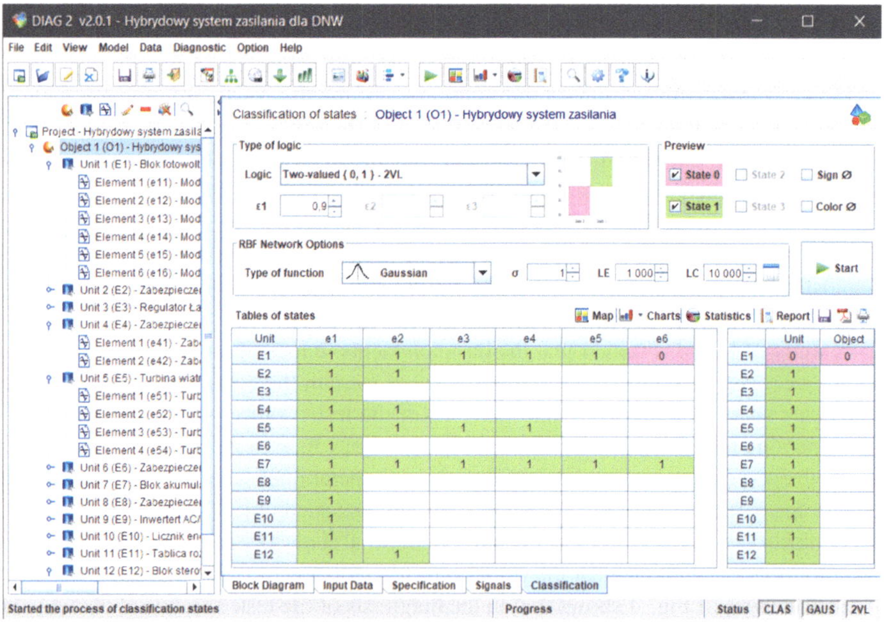

Fig. 3. Resultant form of DIAG 2, “Table of basic element states” in the diagnostic process of the solar power station equipment with 2-valued logic, with: {1} - operational state; {0} - non-operational state.

On the other hand, for the condition of inability, state - “0” one element has been classified. Diagnostics of the same object in 2-value logic classified the fitness status, state - “1” for 27 basic elements, while for the state of inability, state - “0” was classified 1 element. Thus, the sum of all recognized states in an object is 28.

a) b)

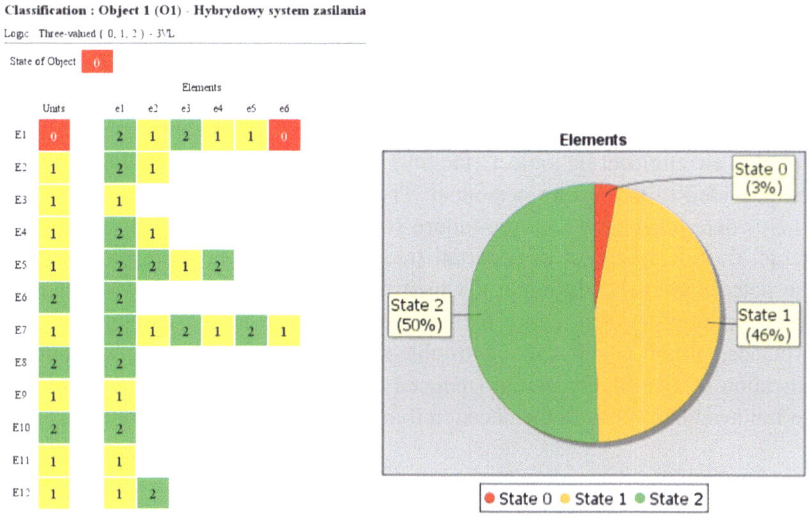

Fig. 4. Resultant form DIAG 2 for the “Classification”, where: (a) is a set of the state of basic elements, (b) is percentage pie chart of basic element states of the object with {2} – operational state; {1} – partial (reduced) operational state; {0} – non-operational state.

To assess the informability of diagnoses determined in 3-value logic in relation to the 2-value logic, the dependence (4) was used. Assuming the following input data: n^{3VL} - number of imperfect latencies, state - “1” is 14, n^{2VL} - number of non-complete states for logic (D^{2VL}) is 0, n^{VL} total number of states determined in the diagnostic process is 28 hence the increase in diagnostic information (D^{3VL}) is:

$$E^{VL} = \frac{n^{3VL} - n^{2VL}}{n^{VL}} \cdot 100\% = \frac{14 - 0}{28} \cdot 100\% = 50\% \quad (5)$$

where: E^{VL} – function (entropy) increment of additional diagnostic information, diagnosis expressed in four-valued logic: n^{3VL} – number of states determined in the diagnostic process, for 3-value logic, n^{2VL} – number of states determined in the diagnostic process for logic (D^{2VL}), n^{VL} – total number of states determined in the diagnostic process.

Based on the obtained results of the diagnosis of the technical object, it has been shown that the use in the diagnosis process in 3-value logic makes this diagnostic more useful. Calculated (E^{VL}) - the entropy (increment) of additional diagnostic information is 50%. Diagnosis expressed in 3-value logic (D^{3VL}) is much more useful for the use of this information in the process of technical service organization than diagnosis (D^{2VL}). Interpretation (recognition) of incomplete airworthiness status, state - "1" in the facility gives the opportunity to apply the optimal organization strategy for technical support (renewal) of the facility based on its condition.

4 Conclusions

The 3-valued logic, unlike 2-valued logic diagnostics, is unique in that its functions and arguments can assume one of three values, each denoted by the symbols from the set {0, 1, 2}. The addition of another reliability (technical) state by 3-valued logic to characterize an engineering object (facility) increases the value of diagnostic information provided to the testing personnel. The additional partial (reduced) operational state helps determine time to a pre-failure state in an engineering facility with better accuracy. The identification of a partial (reduced) operational state provides a more precise determination of the time at which the engineering facility will enter a state which is the most frequent cause for maladjustment, detuning, derating and reduced performance, all of which can be trouble-shot with reconditioning servicing. The identification of one of the partial (reduced) or critical states in the interval of time before failure should prompt the decision to recondition the tested engineering facility.

References

1. Buchannan, B., Shortliffe, E.: Rule – Based Expert Systems. Addison – Wesley, Boston (1985)
2. Duer, S.: Investigation of the operation process of a repairable technical object in an expert servicing system with an artificial neural network. Neural Comput. Appl. **19**(5), 767–774 (2010)
3. Duer, S.: Qualitative evaluation of the regeneration process of a technical object in a maintenance system with an artificial neural network. Neural Comput. Appl. **20**(5), 741–752 (2011)
4. Duer, S.: Modelling of the operation process of repairable technical objects with the use information from an artificial neural network. Exp. Syst. Appl. **38**, 5867–5878 (2011)
5. Duer, S.: Examination of the reliability of a technical object after its regeneration in a maintenance system with an artificial neural network. Neural Comput. Appl. **21**(3), 523–534 (2012)
6. Dhillon, B.S.: Applied Reliability and Quality, Fundamentals, Methods and Procedures. Springer, London (2006)
7. Hayer-Roth, F., Waterman, D., Lenat, D.: Building Expert Systems. Addison - Wesley Publishing Company, Boston (1983)
8. Hojjat, A., Shih-Lin, H.: Machine Learning, Neural Networks, Genetic Algorithms and Fuzzy Systems. Wiley, New Jersey (1995)

9. Kacalak, W., Majewski, M., Zurada, J.M.: Intelligent e-learning systems for evaluation of user's knowledge and skills with efficient information processing. In: Rutkowski, L., Scherer, R., Tadeusiewicz, R., Zadeh, L.A., Zurada, J.M. (eds.) ICAISC 2010. Lecture Notes in Computer Science, vol. 6114, pp. 508–515. Springer, Heidelberg (2010)
10. Lipinski, D., Majewski, M.: System for monitoring and optimization of micro- and nano-machining processes using intelligent voice and visual communication. In: Yin, H., et al. (eds.) Intelligent Data Engineering and Automated Learning, IDEAL 2013. Lecture Notes in Computer Science, vol. 8206, pp. 16–23. Springer, Berlin, Heidelberg (2013)
11. Majewski, M., Kacalak, W.: Smart control of lifting devices using patterns and antipatterns. In: Silhavy, R., et al. (eds.) Artificial Intelligence Trends in Intelligent Systems, CSOC 2017. Advances in Intelligent Systems and Computing, vol. 573, pp. 486–493. Springer, Cham (2017)
12. Majewski, M., Kacalak, W.: Innovative intelligent interaction systems of loader cranes and their human operators. In: Silhavy, R., et al. (eds.) Artificial Intelligence Trends in Intelligent Systems, CSOC 2017. Advances in Intelligent Systems and Computing, vol. 573, pp. 474–485. Springer, Cham (2017)
13. Mathirajan, M., Chandru, V., Sivakumar, A.I.: Heuristic algorithms for scheduling heat-treatment furnaces of steel casting industries. Sadahana **32**(5), 111–119 (2007)
14. Nakagawa, T.: Maintenance Theory of Reliability. Springer, London (2005)
15. Nakagawa, T., Ito, K.: Optimal inspection policies for a storage system with degradation at periodic tests. Math. Comput. Model. **31**, 191–195 (2000)
16. Pokoradi, L.: Logical tree of mathematical modeling. Theory Appl. Math. Comput. Sci. **5**(1), 20–28 (2015)
17. Rosiński, A.: Design of the electronic protection systems with utilization of the method of analysis of reliability structures. In: Proceedings of the Nineteenth International Conference on Systems Engineering ICSEng2008, Las Vegas, USA, pp. 421–426 (2008)
18. Rosiński, A.: Reliability analysis of the electronic protection systems with mixed m–branches reliability structure. In: Berenguer, G., Guedes, S. (eds.) Advances in Safety Reliability and Risk Management. Taylor & Francis, London (2012)
19. Siergiejczyk, M., Krzykowska, K., Rosiński, A., Grieco, L.A.: Reliability and viewpoints of selected ITS system. In: 25th International Conference on Systems Engineering (ICSEng2017), Las Vegas, USA, pp. 141–146. IEEE CPS (2017)
20. Siergiejczyk, M., Paś, J., Rosiński, A.: Issue of reliability–exploitation evaluation of electronic transport systems used in the railway environment with consideration of electromagnetic interference. IET Intell. Transp. Syst. **10**(9), 587–593 (2016)
21. Zajkowski, K.: The method of solution of equations with coefficients that contain measurement errors, using artificial neural network. Neural Comput. Appl. **24**(2), 431–439 (2014)

Use of Light Scattering Method to Assess the Texture of Electropolished Surfaces

Czeslaw Lukianowicz(✉)

Faculty of Mechanical Engineering, Koszalin University of Technology, Raclawicka 15-17, 75-620 Koszalin, Poland
czeslaw.lukianowicz@tu.koszalin.pl

Abstract. The electropolishing process is a frequently used method of finish surface treatment. This process allows to obtain a shiny surface with small surface roughness values. The paper shows that for electropolished surfaces a fast assessment of the surface texture can be made using the light scattering method. The primary aim of this article is to present the theoretical basis of this method, as well as to show the results of its verification by means of experimental research. Samples made of AISI 304 stainless steel were used for the tests. The samples were electropolished in a solution of sulfuric acid and orthophosphoric acid. The solution were prepared in two different concentrations. During the tests, the evaluation of the surface texture of the samples was carried out after different times of electropolishing. The assessment of the geometrical structure of the samples was made using the scatter method and the obtained results were compared with the reference results obtained from surface measurements by the stylus profilometer. The conducted research has shown that the proposed light scatter method can be used to quickly assess the geometric structure of electropolished surfaces.

Keywords: Surface texture · Electropolishing · Light scattering

1 Introduction

Electropolishing is an advanced technique of electrochemical treatment of metals and their alloys. It involves electrochemical removal of material from the surface. This technique is used for machining steel, including stainless steels, copper alloys, aluminium, titanium, etc. The main purpose of electropolishing is to smooth the product surface. The smoother surface is characterized by a smaller value of the developed interfacial area ratio *Sdr* [1]. This means that the reduced real surface area is conducive to reducing of impact on the surface such external factors as, for example, corrosive agents. Furthermore, often the electrochemical polishing process also leads to deep anodic passivation of the surface. As a result, the surface is covered with a passive layer and is largely protected against corrosive agents. In addition, for some types of steel, the electrochemical polishing process promotes surface curing. Moreover, for some types of steel, the electrochemical polishing process promotes surface curing. For example, electropolishing chromium-nickel steel increases the concentration of chromium atoms on the metal surface because iron and nickel undergoes the process of

M. Majewski and W. Kacalak (Eds.): IIRTS 2019, LNME, pp. 148–157, 2020.
https://doi.org/10.1007/978-3-030-37566-9_14

pickling faster than chromium. As a result, the treated surface undergoes additional hardening.

A special feature of the electropolishing process is that the material near the tops of the surface roughness undergoes more intensive chemical dissolution in the electrolyte [2, 3]. This is because during the electropolishing process, a higher current density occurs around the surface roughness peaks. In areas of the surface microcavities, the dissolution process of the chemical surface material is somewhat slower. The result of the electropolishing process is significant smoothing of the product surface and rounding of sharp edges. Properly conducted electropolishing process does not disturb the macroscopic structure of the surface and does not introduce residual surface stresses. The surface roughness decreases and *Ra* and *Sa* parameters reach values below 200 nm [4]. As a result of electrochemical polishing, the surface is smoothed and shiny.

Figure 1 shows the smoothing of the surface during the electrochemical polishing process. The object is immersed in the electrolyte and attached to the anode. The results and efficiency of electrochemical polishing depend on many factors. In regard to electropolishing of stainless steels, the most commonly used electrolytes are solutions of acid sulfuric and orthophosphoric acid. In the case of titanium alloys and other metals, solutions of chromic acid, nitric acid, ammonium hydrogen difluoride and inhibitors are used.

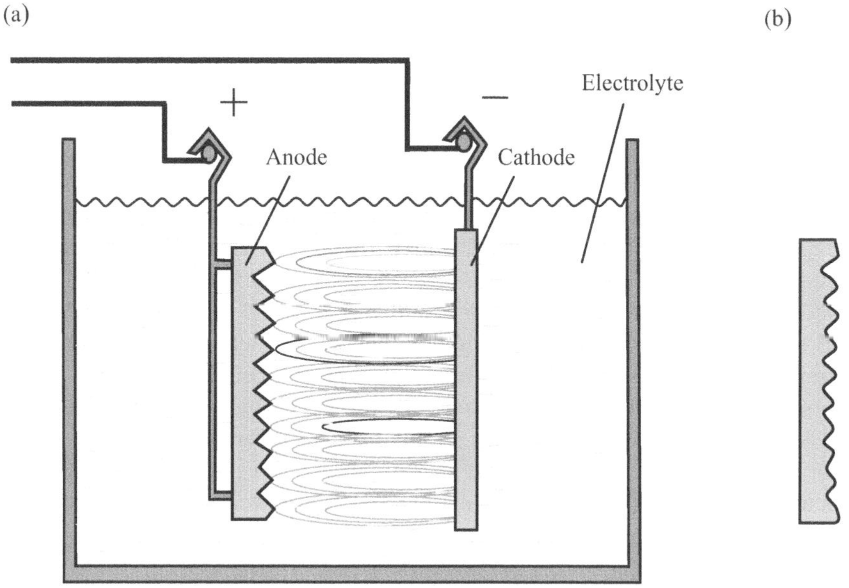

Fig. 1. Smoothing the surface in the electropolishing process (a); the surface profile after electropolishing (b).

2 Texture of Electropolished Surfaces

The surface texture of the electropolished surfaces is generally characterized by the parameters and functions provided for in the standards [5, 6]. Apart from some height parameters such as *Ra*, *Sa*, *Rq* and *Sq*, most often for this purpose so-called hybrid parameters are used [7, 8]. Hybrid parameters include, among others:

Rdq (*RΔq*) - root mean square slope of the assessed profile,
Sdq (*SΔq*) - root mean square gradient of the surface,
Sdr - developed interfacial area ratio,

which are expressed by the following equations:

$$Rdq = \sqrt{\frac{1}{L}\int_0^L \left(\frac{dz}{dx}\right)^2 dx}, \tag{1}$$

$$Sdq = \sqrt{\frac{1}{A}\iint_A \left[\left(\frac{\partial z(x,y)}{\partial x}\right)^2 + \left(\frac{\partial z(x,y)}{\partial y}\right)^2\right] dx\,dy}, \tag{2}$$

$$Sdr = \frac{1}{A}\iint_A \left[\sqrt{1 + \left[\left(\frac{\partial z(x,y)}{\partial x}\right)^2 + \left(\frac{\partial z(x,y)}{\partial y}\right)^2\right]} - 1\right] dx\,dy, \tag{3}$$

where: L is the length of the analyzed profile $z(x)$, $z(x, y)$ is a function describing the uneven surface and A is the surface area of the analyzed region.

The main reason for using hybrid parameters to assess the texture of electropolished surfaces is that these parameters are more sensitive to the surface smoothing process than height parameters such as *Ra*, *Sa*, *Rq* and *Sq*. This is due to the course of the electropolishing process, during which the material near the sharp tops of inequality undergoes to more intense chemical dissolution. Hence, the values of the gradients and the surface derivatives decrease. As a result, according to Eqs. (1), (2) and (3), the values of parameters *Rdq*, *Sdq* and *Sdr* also decrease.

It is worth explaining in more detail why hybrid parameters are more sensitive to changes of the electropolished surface texture than height parameters. For this purpose, we consider a surface whose selected profile is described by the function $z(x)$. Let us analyze the changes taking place in the $z(x)$ profile and its derivative $z'(x)$ as a result of electropolishing. If the spectral power density of the $z(x)$ profile is expressed by a certain function of $S_z(\omega)$, then according to [9] the power spectral density of the derivative of the profile $S_{z'}(\omega)$ is given by the equation:

$$S_{z'}(\omega) = \omega^2 S_z(\omega), \tag{4}$$

where: $\omega = 2\pi f$ is the angular frequency and f is the spatial frequency of the surface profile and its derivative. Unit of measure of the spatial frequency is 1/mm. The above equation shows that even small changes in the height of the surface profile, occurring during the electropolishing process, primarily decrease the harmonic components of the power spectrum of the profile with higher spatial frequencies. These changes, according to the Eq. (4), will significantly affect the spectral power density of the profile derivative, rather than the spectral power density of the surface profile. Therefore, hybrid parameters are generally more sensitive than the height parameters to changes in surface irregularities occurring during electropolishing.

3 Light Scattering Method

The proposed method of assessing the surface texture is a goniophotometric method, which is called angle-resolved scattering and is designated by the acronym ARS. It consists in the measurement and analysis of the angular distribution of light intensity dispersed by the surface under investigation [10]. Among other things, in [11] and [12] it is shown that this distribution is proportional to the probability density function of the slope angles of asperities. In works [12, 13] and [14] it was found that for smooth surfaces the variance Aq of the standardized angular distribution of scattered light intensity $H(\varphi_i')$ correlates with the mean square slope of the roughness profile.

Under the recommendations [12], the parameter Aq is determined on the basis of the n-point distribution of scattered light intensity $I(\varphi_i')$ according to the equation:

$$Aq = \sum_{i=1}^{n} \left(\varphi_i' - M\right)^2 H\left(\varphi_i'\right), \tag{5}$$

where M is the expected value of a standardized distribution of scattering angles. The expected value M is determined from the equation:

$$M = \sum_{i=1}^{n} \varphi_i' H\left(\varphi_i'\right), \tag{6}$$

while $H(\varphi_i')$ is given as

$$H\left(\varphi_i'\right) = \frac{I\left(\varphi_i'\right)}{\sum_{i=1}^{n} I\left(\varphi_i'\right)}, \tag{7}$$

where φ_i' is the tangent of the scattering angle φ_i, i.e. $\varphi' = \tan\varphi$. This method assumes that light scattering is described by the angular distribution of the scattered light intensity $H(\varphi_i')$. In addition, an approximate equation:

$$\varphi' = \tan\,\varphi \cong \varphi, \tag{8}$$

can be used for small scattering angles.

The structure of the system used for practical implementation of the proposed method is shown in Fig. 2(a). In this system, the laser beam is directed perpendicular to the tested surface. The system is used to measure parameter Aq and determine the angular distribution of scattered light intensity, as shown in Fig. 2(b).

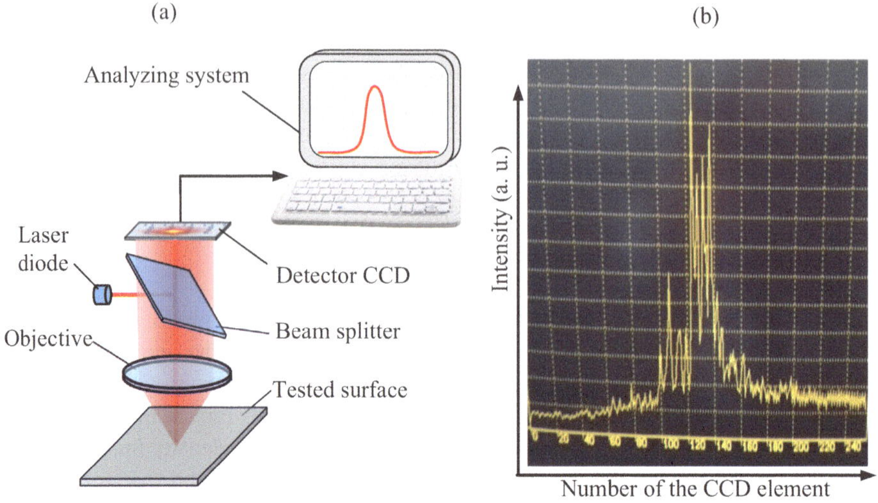

Fig. 2. The system for surface assessment by the method of light scattering (a); distribution of the scattered light intensity (b).

4 Experiments and Results

The aim of the experimental research was to check the suitability of the proposed scatterometric method for quick, non-contact assessment of the surface texture after electropolishing. The tests were carried out on 10 samples made of AISI 304 steel polished electrochemically in a solution of sulfuric acid and orthophosphoric acid. Two anode current densities were used: for samples marked with the EP symbol – 8 A/dm^2 and for samples marked with the symbol HC – 16 A/dm^2. The duration of the electropolishing process was changed in the range from 30 s to 360 s.

For each sample, the distribution of scattered light intensity and the value of parameter *Aq* were determined by the system shown in Fig. 2(a). The measurement system uses a laser diode emitting radiation with a wavelength of 810 nm and a CCD detector containing 256 elements.

The Hommelwerke T8000 profilometer was applied for contact measurements. Using the profilometer, maps and surface images, profiles and roughness parameters were determined, including the *Rdq* hybrid parameter.

The results of optical measurements of the surface texture were compared with the results received by the contact method. A comparison was also made of the results testing the sample surfaces obtained after different electropolishing times. The surface texture for electrochemically polished samples at different anodic current densities was also analyzed. Surface topographies and the roughness parameters of the electropolished samples were also compared with the test results obtained for the unpolished reference sample. The results obtained from the profilometric measurements of the reference sample are shown in Fig. 3. A short list of the results obtained is presented in Table 1. This table contains the values of *Ra*, *Sa*, *Rdq* and *Aq* parameters of the unpolished reference sample and selected samples obtained for different treatment times and anodic current densities.

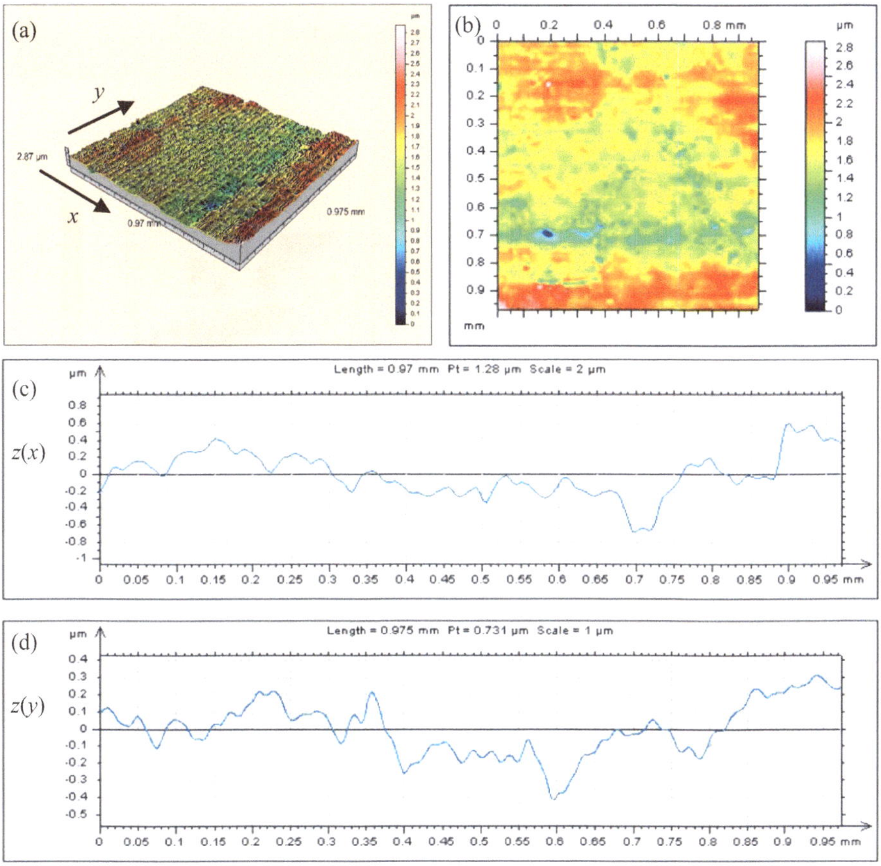

Fig. 3. Results obtained from the profilometric measurements of the reference sample: 3D view of the surface (a); 2D pseudo color map (b); selected surface profile in the x direction (c); selected surface profile in the y direction (d).

Table 1. Results of surface roughness measurements of electropolished samples.

Sample	Time (s)	*Ra* (µm)	*Sa* (µm)	*Rdq* (deg)	*Aq*
Reference	0	0,136	0,345	0,497	74
EP30s	30	0,121	0,339	0,488	63
EP3m	180	0,115	0,296	0,366	48
EP6m	360	0,052	0,179	0,325	15
HC30s	30	0,081	0,252	0,290	32
HC3m	180	0,094	0,300	0,271	27
HC6m	360	0,067	1,700	0,186	16

Figure 4 shows 2D pseudo-color surface maps of the electropolished samples obtained from the profilometer. It can be seen that the surface of the HC6m sample, shown in Fig. 4(f), has been significantly changed due to the longer duration of the electropolishing process and a higher anode current density. This may mean that after a certain time of electropolishing the pitting process started.

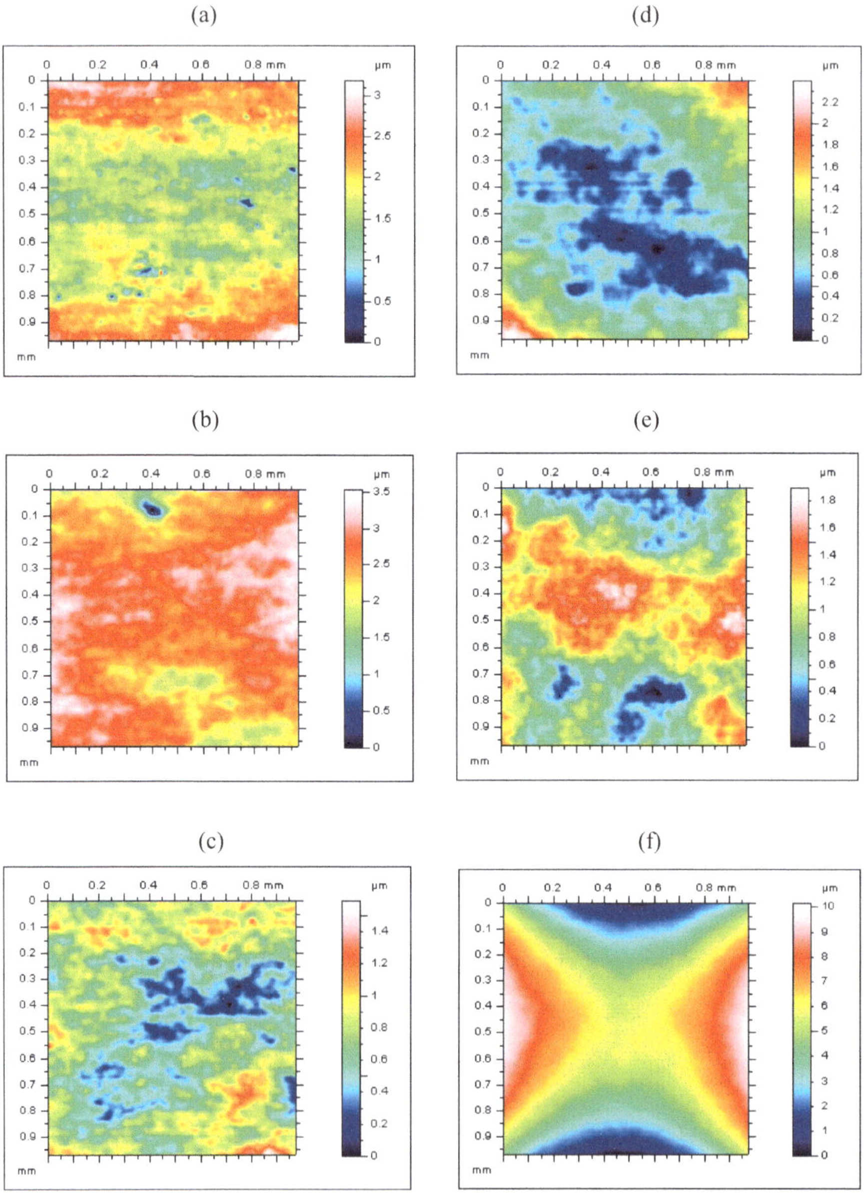

Fig. 4. 2D pseudo-color surface maps of the electropolished samples; sample EP30s (a); sample EP3m (b); sample EP6m (c); sample HC30s (d); sample HC3m (e); sample HC6m (f).

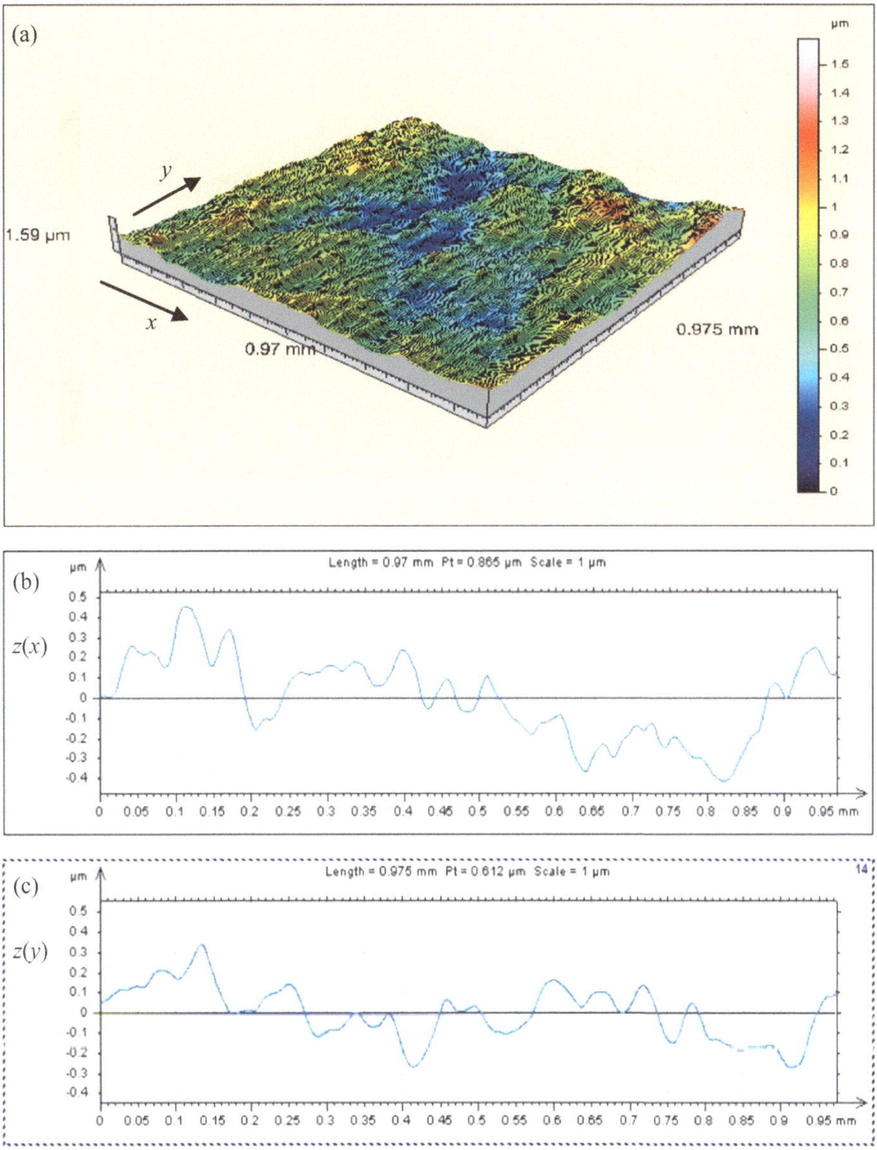

Fig. 5. Results of profilometric measurements of the EP6m sample: 3D view of the surface (a); selected surface profile in the x direction (b); selected surface profile in the y direction (c).

Due to the lower value of the anodic current density, this process did not occur or was significantly weaker during the electropolishing process of the EP6m sample. This is due to the comparison between Figs. 5 and 6. It is worth noting that in the case of the HC6m sample, significant changes in the shape of the surface occurred in a relatively short time. Regardless, the surface remained smooth and shiny.

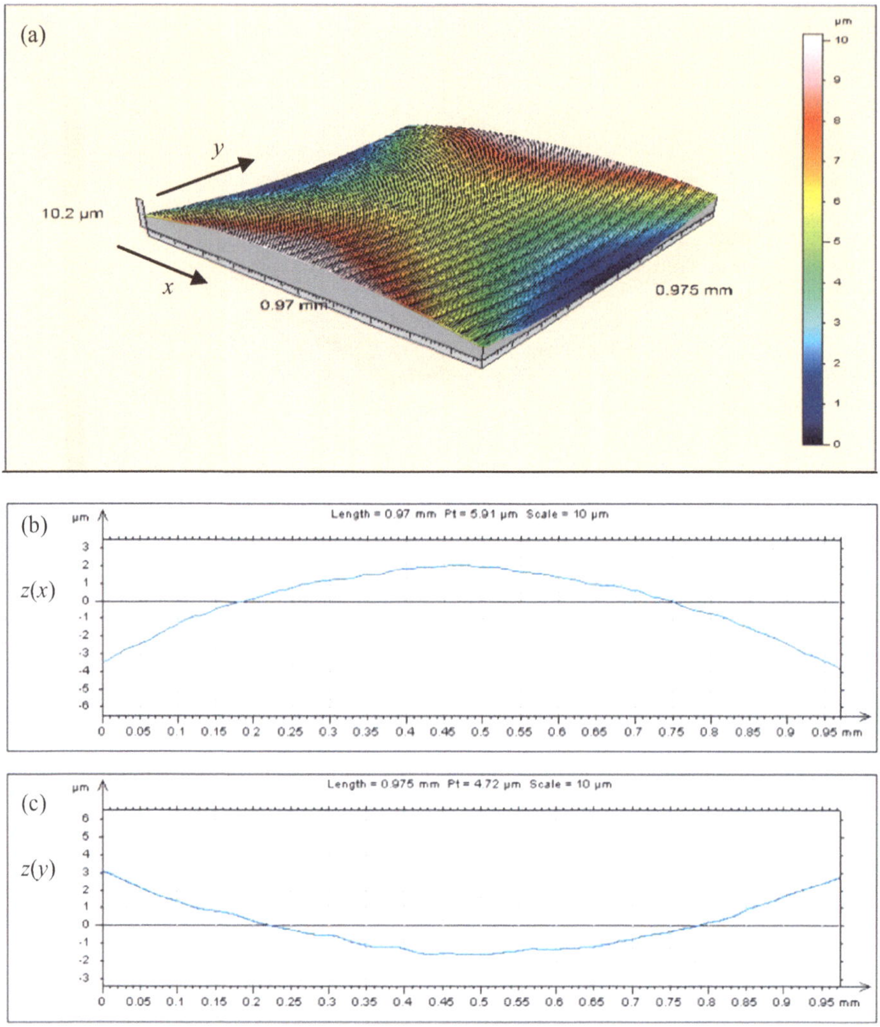

Fig. 6. Results of profilometric measurements of the HC6m sample: 3D view of the surface (a); selected surface profile in the x direction (b); selected surface profile in the y direction (c).

5 Conclusions

The process of shaping the surface texture during electropolishing was analyzed. It has been shown that hybrid parameters are useful for assessing the surface roughness. A scatterometric method for fast assessment of the electropolished surfaces has been proposed. This method allows to assess the surface roughness based on measurements of the parameter Aq being a variance of the scattering angles of light reflected from the surface. The experiments confirmed that the values of the Aq parameter are correlated with the values of the Rdq parameter.

Based on the analysis of the presented results, it can be concluded that the proposed scatterometric method can be useful in industrial applications for quick, non-contact assessment of the surface texture after electropolishing.

The proposed method can also be used to assess the geometric structure of surfaces finished by other methods. In order to make greater use of the proposed method, further research is planned. One of the directions of such works may be the adaptation of this method to monitoring the processes of precision surface treatment.

References

1. Lochynski, P., Sikora, A., Szczygiel, B.: Surface morphology and passive film composition after pickling and electropolishing. Surf. Eng. **5**(33), 395–403 (2017)
2. Hryniewicz, T.: Concept of microsmoothing in the electropolishing process. Surf. Coat. Technol. **2**(64), 75–80 (1994)
3. Hryniewicz, T., Rokicki, R., Rokosz, K.: Surface characterization of AISI 316L biomaterials obtained by electropolishing in a magnetic field. Surf. Coat. Technol. **9**(202), 1668–1673 (2008)
4. Núñez, P.J., García-Plaza, E., Hernando, M., Trujillo, R.: Characterization of surface finish of electropolished stainless steel AISI 316L with varying electrolyte concentrations. Proc. Eng. **63**, 771–778 (2013)
5. EN ISO 4287:1998/A1:2009, Geometrical Product Specifications (GPS) – Surface texture: Profile method – Terms, definitions and surface texture parameters – Amendment 1: Peak count number
6. EN ISO 25178-2:2012, Geometrical product specifications (GPS) – Surface texture: Areal – Part 2: Terms, definitions and surface texture parameters
7. Nakar, D., Feuermann, D.: Surface roughness impact on the heat loss of solar vacuum heat collector elements (HCE). Renew. Energy **96**(Part A), 148–156 (2016)
8. Nakar, D., Harel, D., Hirsch, B.: Electropolishing effect on roughness metrics of ground stainless steel: a length scale study. Surf. Topogr.: Metrol. Prop. **1**(6), 015003 (2018)
9. Hajek, B.: Random Processes for Engineers. Cambridge University Press, Cambridge UK (2015)
10. Li, H., Torrance, K. E.: An experimental study of the correlation between surface roughness and light scattering for rough metallic surfaces. In: Duparré, A., Singh, B., Gu, Z.-H. (eds.) Proceedings SPIE, vol. 5878, pp. 58780 V-1–15. SPIE, Bellingham, WA (2005)
11. Geisler, T., Kolb, A.: Pattern recognition of rough surfaces by using goniometric scattered light. Metrol. Meas. Syst. **1**(25), 33–46 (2018)
12. VDA: 2010-07-00, Geometrische Produktspezifikation Oberflächenbeschaffenheit Winkelaufgelöste Streulichtmesstechnik Definition, Kenngrößen und Anwendung (2009)
13. Brodmann, R., Seewig, J.: Non-contact surface metrology by means of light scattering. In: Wang, Q.J., Chung, Y.-W. (eds.) Encyclopedia of Tribology, pp. 2433–2440. Springer, Boston (2013)
14. Feidenhans'l, N.A., Hansen, P.E., Pilný, L., Madsen, M.H., Bissacco, G., Petersen, J.C., Taboryski, R.: Comparison of optical methods for surface roughness characterization. Meas. Sci. Technol. **26**, 085208, 10 p (2015)

Human Image Recognition System with an Artificial Neural Network Working in Three-Valued Logic

Stanisław Duer(✉) and Konrad Zajkowski

Koszalin University of Technology, Raclawicka 15-17, 75-620 Koszalin, Poland
stanislaw.duer@tu.koszalin.pl

Abstract. The paper presents a method for a biometric image recognition of the human using the vector distance measure. The basis of the proposed method of recognition is a biometric characteristic picture of a man (the image of the eye, hands, etc.). To this end, the classification structure was presented of the biometric image. An important part of this study is to present the theoretical basis of the functioning of the proposed pattern recognition method and the neural network operation algorithm. The neural network engineered is a convenient research tool used in the image analysis. The structure of the image recognition system was illustrated and described. An important part of this system is a module for determining the coordinates of singular points (the image scanning module) and the recognition module that uses a SBM type neural network known as the Simile-Based Methods (SBM), working in three-valued logic.

Keywords: Artificial intelligence · Biometrics · Knowledge base · Neural networks · Pattern recognition · Security access control

1 Introduction

The literature is broadly supportive of intelligent systems. They are used in the supervision and organization of technical and technological processes, in terms of their diagnosis and the assessment of quality. These systems are particularly useful when there are many variables analysed and the need to take into account complex factors acting on them, and on: the technical condition of technical equipment, tools, means of production and the selection of conditions and process parameters. Important achievements make the following works [2–5, 11, 16–22].

They developed concepts and detailed systems composed of many specialized modules mainly concern intelligent, comprehensive monitoring system processes, including processes and equipment.

These include the development of Hojjat, Shih–Lin, Kacalak and et al. [2–9, 12–17], which describes the basis for the application of intelligent systems to solve organizational processes and supporting various types of technical and technological problems. An important achievement presented in this work is a description of the transformation of human knowledge - specialist knowledge expressed in the form of computer programming language. This group of studies also include designer work [1, 14–22].

M. Majewski and W. Kacalak (Eds.): IIRTS 2019, LNME, pp. 158–170, 2020.
https://doi.org/10.1007/978-3-030-37566-9_15

In the recent years, in connection with numerous terrorist activities in the world, it is particularly important to increase all types of safety systems to recognize humans in aviation and other types of transport. These systems include various types of circuits and systems for image recognition. General theoretical methods of pattern recognition theory describing have been thoroughly identified, yet those systems and devices that implement them are constantly being improved.

The unique body features such as fingerprints, the palm shape, the face shape, or the pattern of the iris of the eye currently constitute the most effective identification factor (as opposed to e.g. a card, a string, alphanumeric names/passwords/or PIN). Of all the biometric methods, the highest level of security is offered by iris identification. Information on the characteristics of the iris is included in a 512-byte machine code, so that in the process of identification, it is possible to compare the code user with all the other codes stored in the database. The iris is a part of the human body which, owing to its uniqueness, allows the use of the optimal method of electronic identification. It is characterized by a very high accuracy: the iris pattern is complex and unique; the probability of two identical patterns is 1 in 1078 (the risk of error is 1 to 1.2 million).

The problem of researching the process of human recognition presented in the article based on its biometric image of the "eye" or "hand" is developed in two directions.

The first direction of research development is related to the improvement of the diagnostic system of intelligent human recognizer. To this end, proposing rules are developed that allow to develop diagnoses in a four-valued logic. It is assumed that the four-valued logic contains greater possibilities for more accurate human recognition than the trivalent logic currently used for this purpose.

The second direction of research will be related to the theoretical and practical development of a more accurate description of the human biometric image vector. For the purposes of these studies, work is being done in the field of vector reading (description and analysis) of the human biometric image. It is assumed that increasing the module of the human biometric image description vector directly in the fluid to increase the quality of human recognition and the intelligent system.

2 Description of the Biometric Image of the Human

New methods of the recognition of "the humans' personal qualities" are based on pattern recognition methods. The method proposed of assessing the compliance with the individual's image of the eye or the palm is very suitable for this purpose. It can be seen in Fig. 1 that the image of the eye and the palm constitutes a collection of a large number of points: the "personal points" that are specific to that image. Personal points form in a given image "cluster points": collections of clusters with shapes that can be approximated (described) with a line or a geometric shape.

The problems that occur here include the following:

- identification of personal points in the image of the human being recognized,
- adoption of the coordinate system and the measurement vectors of coordinates of the personal points,

- analysis of the measure vectors of the personal point coordinates,
- method of assessing compliance with the human image.

Among the possible coordinate systems that describe the personal characteristics in the image of the face, eyes or palms, or any other body parts, coordinate systems can be used that are related to the following:

- the beginning of the coordinate system in the geometric centre of the test image (Fig. 1),
- the coordinate system such as an external reference (Fig. 1),
- the coordinate system to the selected point trailing the personal image of the test.

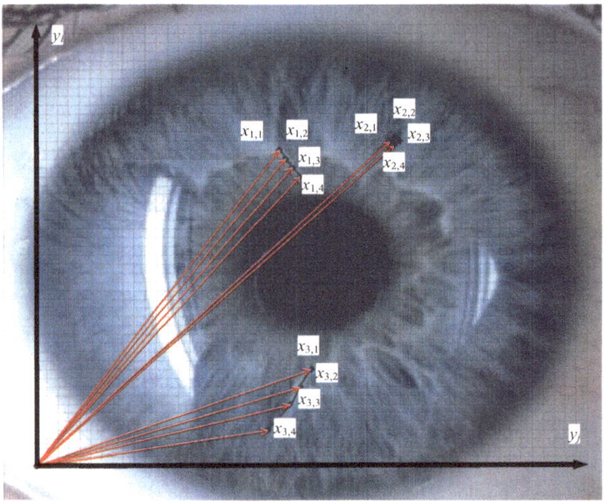

Fig. 1. The coordinates of singular points in the image of the human biometric personal qualities on example of description of the eye image.

The human image recognized (Fig. 1) must be specified to its personal qualities that are specific to a given image; for the palm image, these are characteristic lines that are designated by a specific collection of "personal points". The image of the eye is a collection of "personal points" that make up a certain characteristic "cluster of personal points" that can be described with a line or a geometrical figure. In the analysis process of personal information, it is convenient to describe it as a measure of the vector for the line and the matrix (array) for the cluster. The coordinate vector form of the line of measurement can be presented in the following form:

$$[X(A)] = \lfloor X(A_{1,1}), X(A_{1,2}), \ldots, X(A_{i,j}), \ldots, X(A_{I,J}) \rfloor = \lfloor X(A_{i,j}) \rfloor \quad (1)$$

where: $[X(A_{i,j})]$ – the measure vector coordinate of the i-th image for the j-th singular point that is included on the i-th singular line of this image.

where: $[X(A_{i,j})]$ - a-coordinate of the vector of the a-th image for the j-th point that is located on the i-th singular line of this picture A.

A complete setting of the measure vectors that describe the coordinates of singular points in an image is possible with the table that is used for this purpose (cf. Table 1).

Table 1. Table of input measurement vectors of the personal points of a human.

Picture of a human	Singular line of this picture A	Vectors of the personal points of a human $\{A_{i,j}\}$				
A	E_1	$X(A_{1,1})$	...	$X(A_{1,j})$	...	$X(A_{1,J})$
	⋮	⋮	...	⋮	...	⋮
	E_i	$X(A_{i,1})$	...	$X(A_{i,j})$	...	$X(A_{i,J})$
	⋮	⋮	...	⋮	...	⋮
	E_I	$X(A_{I,1})$	...	$X(A_{I,j})$	...	$X(A_{I,J})$

2.1 The Algorithm and the Structure of the DIAG Intelligent Computer Program in the Image Recognition System

The objective of the neural network that is used in the pattern recognition system is a comparison of the vectors of the coordinates of the image singular points with the coordinates of the standard (nominal) points. For this purpose, it is convenient to present the coordinates of the image singular points in a vector form. The analytical form of the equation that describes the process of comparing the images with their standard images is presented as follows [2–5]:

$$\underset{A_{i,j}\in\{E_i\}}{\forall} \exists\left(\lfloor X(A_{i,j})\rfloor \mapsto \left\lfloor X\left(A_{(w)_{i,j}}\right)\right\rfloor\right) \Rightarrow \lfloor W(\varepsilon(A_{i,j}))\rfloor \tag{2}$$

where: $[X(A(w)_{i,j})]$ – the vector of the coordinates of the reference image singular point that is located at the j-th point and on the i-th line, $[X(A_{i,j})]$ - the vector of coordinates of the j-th singular point located on the i-th line of the image, $\lfloor W(\varepsilon(A_{i,j}))\rfloor$ - the resultant vector of the diagnosis (reference) of the j-th point on the i-th line of the image - the universal quantifier - the existential quantifier - the relation of comparison, - the relation of entailment.

Relation (1) shows a description of the pattern recognition (diagnosing) method (Fig. 2). The relationship demonstrates that for each j-th point on the i-th line of the image, there exists an image point $[A_{i,j}]$ that is described as a coordinate vector, which is compared with its vector image of the benchmark. The result of this diagnostic action is the determined compliance status of the diagnosis of this point $(\varepsilon(A_{i,j}))$. Thus, diagnosing actions involving a comparison of the coordinate vectors of the image singular points with the coordinates of the vectors of their calibration is performed by the diagnosing subsystem. This task becomes more effective in converting information if an artificial neural network is used in this process. The network is to develop a decision: to identify to which of the three distinguished classes of the diagnosis compliance status {2, 1, 0} a given state of the image singular point is to qualify. The image recognition (comparison) method as presented in this study is in general a representation of a wider theory known in the literature of as the pattern recognition theory covered by the papers [5–15].

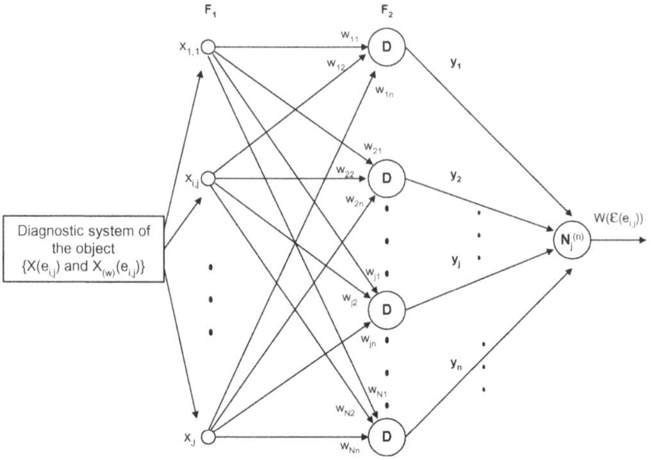

Fig. 2. Structure of artificial neural network system DIAG.

In the structure of the image recognition algorithm – the operation program of the network presented in Fig. 2, after an initial conversion of the image information signal, a signal is generated which is fed to all the cells in the form of the following relationship:

$$f_i(x, w) = \sum_{i=1}^{K} \omega_{i,j} \cdot X_i \tag{3}$$

The results of the image (object) recognition test according to the diagram in Fig. 2, which are calculated as a result of relations (1 to 2), are presented in a table of states (Table 2).

Table 2. Table of the state of the object's items.

Vector of the state of the object's items $\{e_{i,j}\}$				
$\varepsilon(A_{1,1})$	…	$\varepsilon(A_{i,j})$	…	$\varepsilon(A_{i,J})$
$W(\varepsilon(A_{1,1}))$	…	$W(\varepsilon(A_{1,j}))$	…	$W(\varepsilon(A_{1,J}))$
⋮	…	⋮	…	⋮
$W(\varepsilon(A_{i,1}))$	…	$W(\varepsilon(A_{i,j}))$	…	∅
⋮	…	⋮	…	⋮
$W(\varepsilon(A_{I,1}))$	…	$W(\varepsilon(A_{I,j}))$	…	$W(\varepsilon(A_{I,J}))$

where: $W(\varepsilon(A_{i,j}))$ – recognize the value of the j-th singular point located in the i-th line of the image, expressed in values from the set of the adopted three-valued logic state assessment – {2, 1, 0}), ∅ – element complementary dimension of the array.

3 Investigations into the Compliance of the Recognition of the Human Personal Characteristics in a Biometric Image Using an Artificial Neural Network Working in the Three-Valued Logic

In order to verify the proposed method of the conformity assessment of the human personal characteristics with the use of an artificial neural network, the human palms were chosen as a biometric image (Fig. 3). In the palm image, characteristic lines were selected that were designated (described) by singular points, which are summarized in Table 1.

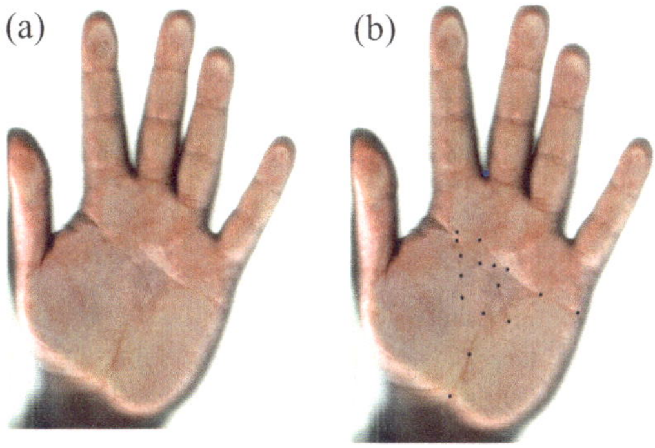

Fig. 3. Biometric palm image, where: (a) image of the palm with no singular points, (b) image of the palm including singular points.

Considering a simplification of the calculations, the number of the i-th lines of the palm-in-picture was reduced to four, and the number of the j-th singular points on the line was limited to five. In practice, the number of lines and singular points in the image can be infinitely large. Studies have shown that the pressure of the palm during the scan has a significant impact on the quality of the image recognition in the method proposed. The pressure of the palm during the scan cannot be too strong or too weak. When scanning an image with different palm pressure forces, it became evident that the coordinates that describe the singular points of the scanned palm image are very different from the reference (real) image of the palm. This problem was solved by choosing a singular point in the test image. This point is known as the ***point of correction***.

The correction point in the image tested is considered to be a singular point of the image, whose coordinates (X and Y) are fixed and invariable regardless of the palm contact force during the scan.

After scanning of the palm image, the image is graphically processed using any graphics program available. This allows one to designate the "map of the palm" of coordinate vectors (Fig. 4).

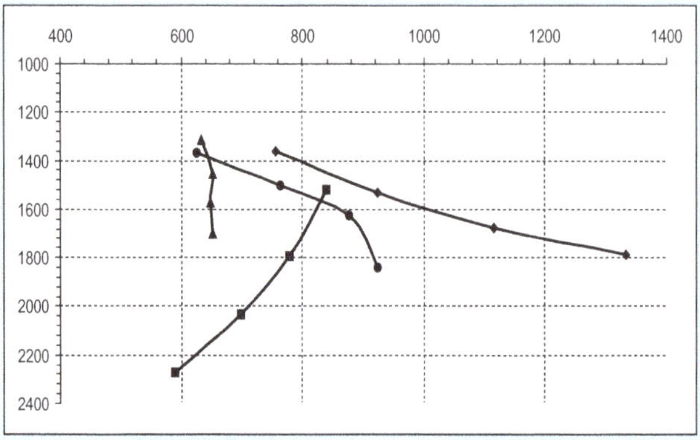

Fig. 4. Coordinates of the biometric palm image: the "matrix of singular points of the image – the image bitmap".

In the map of this "image matrix", the coordinates (X and Y) of the bitmap are assigned to the highlighted singular points, and the coordinates of the correction point CP (A) are defined. The removal of the coordinates of the singular points is performed twice; this especially relates to the CP coordinates. Further, an analysis is performed of the compatibility comparisons of the results of measurement 1 and measurement 2. If there is a difference between the coordinates that describe the location of the CP in the bitmap image, the coordinates (X and Y) of the singular points need to be recalculated taking into account the coordinate corrective patch:

$$\left(\mathrm{CP}\left(A_{1X,Y}\right) - \mathrm{PK}\left(A_{2X,Y}\right) = \Delta\mathrm{PK}\left(A_{X,Y}\right)\right) \tag{4}$$

The vector of the coordinates of the reference point (that corrects the vectors of the coordinates of the singular points of the line) possesses parameters (792, 976). The performance of the task that corrects the coordinates of the bitmap image singular points tie the coordinates of these points to the CP point coordinates. The coordinates of the singular points of the image after the correction recalculation are presented in Tables 3 and 4. The coordinates included in Tables 3 and 4 are at the same time the coordinates of the learning vectors of the network.

Table 3. An array of singular point vectors of the image for coordinate x.

Picture	Line of the picture	Vector learner (X) of singular points of the test image [X_j]			
A	E_1	757	923	1114	1332
	E_2	626	764	876	923
	E_3	632	651	648	651
	E_4	839	779	698	589

In order to identify the compatibility of the biometric image with its model, an artificial neural network of the DIAG program was used. The results of the comparison of the images are presented in Tables 3 and 4. Figure 5 presents the preliminary result of the operation of the neural network for the step computations of the network selected from the bar in the DIAG "N-metrics".

Table 4. An array of singular point vectors of the image for coordinate y

Picture	Line of the picture	Vector learner (Y) of singular points of the test image [Y_j]			
A	E_1	1364	1532	1679	1785
	E_2	1370	1504	1623	1839
	E_3	1314	1454	1573	1701
	E_4	1517	1795	2032	2270

The results of this phase of the network operation are the calculated differential vectors of metric distances between the image singular points and the points in the reference image.

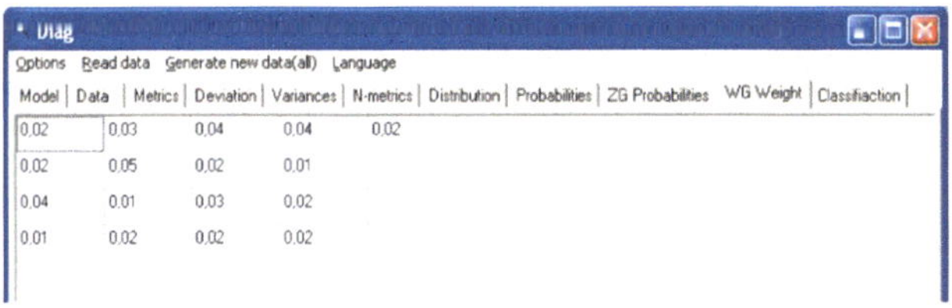

Fig. 5. Display of the calculation results in a neural network in the DIAG program for "N-metrics".

Figure 6 presents an intermediate result of the operation of the neural network in the DIAG program as selected from the "WG Weight" bar. The results of this phase are the calculated weighting factors of the network as an output function.

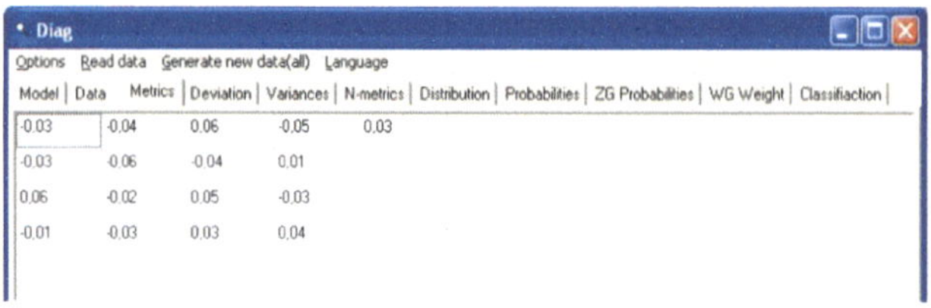

Fig. 6. Screen of calculation results of the neural network in the DIAG program for "WG Weight".

The final result of the work of the artificial neural network of the DIAG program is presented in Fig. 9. The results of this operation phase of the network are the designated (identified) compatibility states of the singular points of the biometric image. The recognized compliance states of the biometric image singular points are presented in three-valued logic, where the individual states mean the following: {2} – the recognition compliance status for the j-th point image (a line or an image), {1} – the state of incomplete recognition compliance for the j-th point image (a line or an image), {0} – the state of non-recognition for the j-th point images (a line or an image).

The table of the recognized j-th singular points (Fig. 7) constitutes the basis for the diagnosing system that generates a decision on recognizing the image with its reference image. The system analyses all the states of the j-th points contained in the i-th lines of the image according to Relation (2). In the absence of any singular points of the image in the "bitmap", as in the case of the "table of the image conformity states", symbol ∅ is inserted meaning that in this field, the array element: the image singular point does not occur. This is only a complementation of the full dimension of the information set. For the information contained in Fig. 7, the system that recognizes the biometric image has developed ***the state {2} that denotes the conformity of the image recognized with its reference image***.

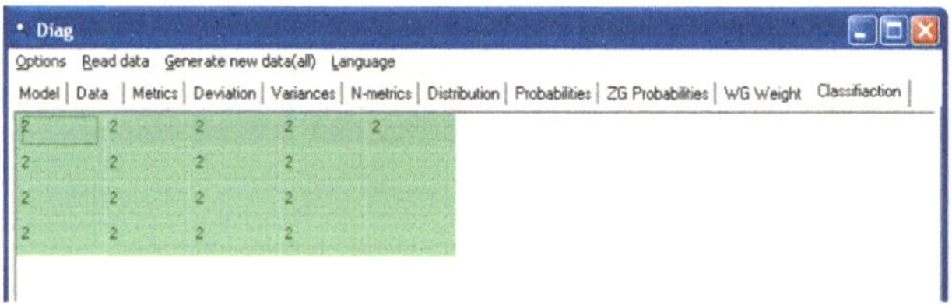

Fig. 7. The screen of the final calculation results of the neural network in the DIAG program.

The work of the system that recognizes the compatibility of the image with its reference image was also verified in the absence of any compliance of the biometric image with its reference image. The calculation results of the neural network for this situation are presented in Figs. 8, 9 and 10.

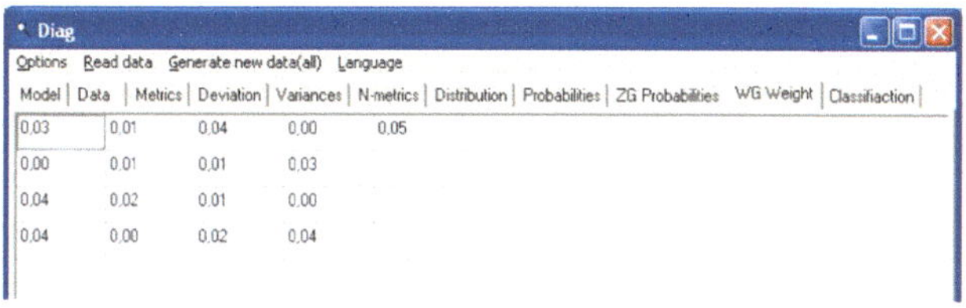

Fig. 8. The screen of calculation results in a neural network in the DIAG program for the type of "N-metrics" (the case of non-compliance with the original picture).

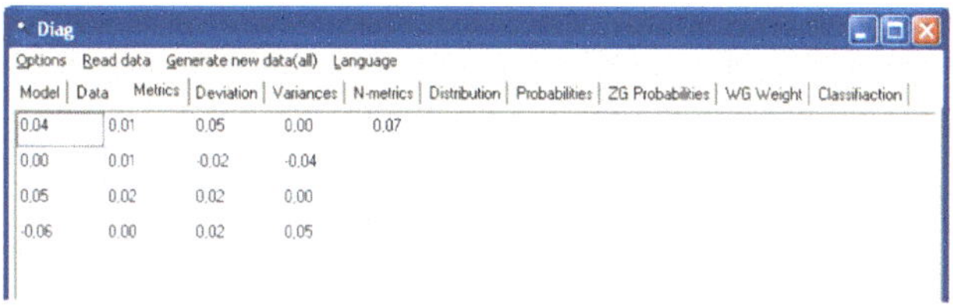

Fig. 9. The screen of calculation results in a neural network in the DIAG program for the type of "WG Weight" (the case of non-compliance with the original picture).

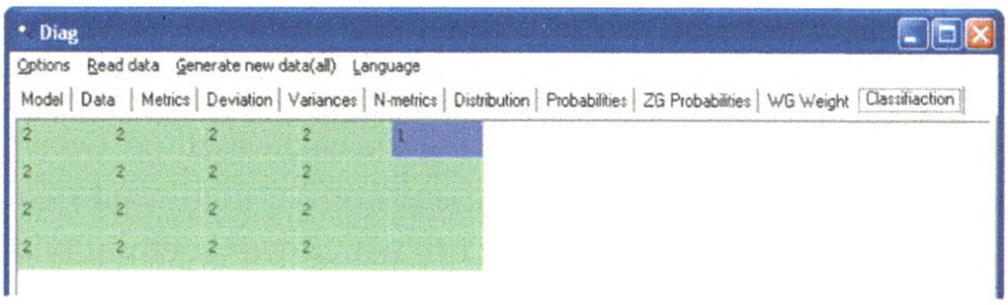

Fig. 10. The screen of final results of the network of neurons in the DIAG program information compiled in the form of the "Compliance status table of the of the image singular points" (the case of non-compliance).

For the results of the operation of the neural network contained in Fig. 10, the system that recognizes a biometric image of the ratios (2) generated a negative decision concerning the recognition of the non-compliance of the image with its reference image. This decision is ***the state {1} denotes the non-compliance of the recognized image with its reference correct image***.

For the results of the operation of the neural network contained in Fig. 10, the system that recognizes a biometric image of the ratios (2) generated a negative decision concerning the recognition of the non-compliance of the image with its reference image. This decision is ***the state {1} denotes the non-compliance of the recognized image with its reference correct image***. Based on the information contained in Fig. 10, it is evident that in the 1st line, singular point ($A_{1,5}$) was recognized with state {1} that represents the state of incomplete compliance of the the j-th point images.

The method presented in this study concerning image recognition that defines (uses) image similarity measures of the SBM type belongs to a group of methods that use vector distance measures. The metric measures of the distance measurement are well suited for the human biometric image recognition. The studies on this subject cover a number of issues that significantly affect the quality of image diagnosis. Among the issues that could materially affect the outcome (decision) concerning the recognition or non-recognition of an image include the image recognition method used, the palm contact force during the scan, the software tool used, the selection of singular points and singular lines in the image, the selection of the weighting in the test image, the decision-making method, etc. Thus, it may be concluded that the problem of the safety of identification of persons on the basis of a biometric image is a difficult organizational and technical undertaking.

4 Conclusions

These days, with a high risk of terrorism, the issue of human image recognition is of particular importance. The security of the identification of persons on the basis of their respective biometric images is an important issue in airport security, banking security etc. In order to identify people based on their biometric images, various methods and techniques of recognition are used. There are as many of these methods as there are known or used methods of pattern recognition.

The paper presents a method of image recognition that uses a vector-metric measure of distance. The method is in some sense a classical method because it is one of the first pattern recognition methods, which the primitive man used (learning to know the world and the surrounding reality) in their life which was seemingly quite primitive. The picture was even then of an importance to this man; the man learned to live on the basis of an image and studied new phenomena. The techniques and methods of pattern recognition developed over the ages.

An important part of the paper is the presentation of the theoretical basis of the functioning of this method and the algorithm of the network. The neural network designed is a convenient research tool used in the analysis of images. There are no such solutions in the literature. They represent a kind of secret considering security reasons in the lives of individuals and societies, such as airports, seaports, etc.

The biometric image recognition method proposed is simple to use, yet it is quite effective. The results obtained confirm the relatively high safety of its use in various practical aspects. This method can be used to identify various forms of images that are presented by the sets of singular points of the image. There is a free selection of the number of singular points of an image. The main problem in this method is the quality of the determination of the coordinates of singular points in the image.

References

1. Buchannan, B., Shortliffe, E.: Rule – Based Expert Systems. Addison – Wesley Publishing Company, Boston (1985)

2. Duer, S.: Diagnostic system with an artificial neural network in diagnostics of an analogue technical object. Neural Comput. Appl. **19**(1), 55–60 (2010)
3. Duer, S.: Diagnostic system for the diagnosis of a reparable technical object, with the use of an artificial neural network of RBF type. Neural Comput. Appl. **19**(5), 691–700 (2010)
4. Duer, S., Duer, R.: Diagnostic system with an artificial neural network which determines a diagnostic information for the servicing of a reparable technical object. Neural Comput. Appl. **19**(5), 755–766 (2010)
5. Duer, S.: Artificial neural network in the control process of object's states basis for organization of a servicing system of technical objects. Neural Comput. Appl. **21**(1), 153–160 (2012)
6. Duer, S., Zajkowski, K., Płocha, I., Duer, R.: Training of an artificial neural network in the diagnostic system of a technical object. Neural Comput. Appl. **22**(7–8), 1581–1590 (2013)
7. Duer, S., Zajkowski, K.: Taking decisions in the expert intelligent system to support maintenance of a technical object on the basis information from an artificial neural network. Neural Comput. Appl. **23**(7–8), 2185–2197 (2013)
8. Duer, S., Zajkowski, K., Duer, R., Paś, J.: Designing of an effective structure of system for the maintenance of a technical object with the using information from an artificial neural network. Neural Comput. Appl. **23**(3–4), 913–925 (2013)
9. Dhillon, B.S.: Applied Reliability and Quality, Fundamentals, Methods and Procedures. Springer, London (2006)
10. Hayer-Roth, F., Waterman, D., Lenat, D.: Building Expert Systems. Addison – Wesley Publishing Company, Boston (1983)
11. Hojjat, A., Shih-Lin, H.: Machine Learning, Neural Networks, Genetic Algorithms and Fuzzy Systems. Wiley, New Jersey (1995)
12. Kacalak, W., Majewski, M., Zurada, J.M.: Intelligent e-learning systems for evaluation of user's knowledge and skills with efficient information processing. In: Rutkowski, L., Scherer, R., Tadeusiewicz, R., Zadeh, L.A., Zurada, J.M. (eds.) ICAISC 2010. Lecture Notes in Computer Science, vol. 6114, pp. 508–515. Springer, Berlin (2010)
13. Lipiński, D., Kacalak, W., Tomkowski, R.: Methodology of evaluation of abrasive tool wear with the use of laser scanning microscopy. Scanning **36**(1), 53–63 (2014)
14. Gupta, M., Liang, J., Noriyasu, H.: Static and Dynamic Neural Networks, from Fundamentals to Advanced Theory. Wiley, New Jersey (2003)
15. Majewski, M., Kacalak, W.: Smart control of lifting devices using patterns and antipatterns. In: Silhavy, R., et al. (eds.) Artificial Intelligence Trends in Intelligent Systems, CSOC 2017. Advances in Intelligent Systems and Computing, vol. 573, pp. 486–493. Springer, Cham (2017)
16. Majewski, M., Kacalak, W.: Innovative intelligent interaction systems of loader cranes and their human operators. In: Silhavy, R., et al. (eds.) Artificial Intelligence Trends in Intelligent Systems, CSOC 2017. Advances in Intelligent Systems and Computing, vol. 573, pp. 474–485. Springer, Cham (2017)
17. Rosiński, A.: Reliability analysis of the electronic protection systems with mixed m–branches reliability structure. In: Berenguer, G., Guedes, S. (eds.) Advances in Safety Reliability and Risk Management. Taylor & Francis, London (2012)
18. Wiliams, J.M., Zipser, D.: A learning algorithm for continually running fully recurrent neural networks. Neural Comput. **1**(2), 270–280 (1989)
19. Zajkowski, K.: The method of solution of equations with coefficients that contain measurement errors, using artificial neural network. Neural Comput. Appl. **24**(2), 431–439 (2014)

20. Zajkowski, K.: Settlement of reactive power compensation in the light of white certificates. In: E3S Web of Conferences, vol. 19, no. 01037 (2017). https://doi.org/10.1051/e3sconf/20171901037
21. Zajkowski, K.: An innovative hybrid insulation switch to enable/disable electrical loads without overvoltages. In: E3S Web of Conferences, vol. 19, p. 01033 (2017). https://doi.org/10.1051/e3sconf/20171901033
22. Zajkowski, K., Rusica, I., Palkova, Z.: The use of CPC theory for energy description of two nonlinear receivers. In: MATEC Web of Conferences, vol. 178, p. 09008 (2018). https://doi.org/10.1051/matecconf/201817809008

Teleoperation Control System for Controlling Prototype of a Loader Crane

Marcin Woźniak(✉), Paweł Herbin(✉), and Arkadiusz Parus(✉)

West Pomeranian University of Technology Szczecin,
17 Piastów Ave., 70-310 Szczecin, Poland
{Marcin.Wozniak, Pawel.Herbin,
Arkadiusz.Parus}@zut.edu.pl

Abstract. The paper presents the results of two different control systems for controlling a loader crane in the Master - Slave configuration: 1st configuration is feedforward and feedback, 2nd configuration is feedback. The control object is an upper limb exoskeleton, which is controlled by human muscles by the contact force, which is measured on tensometric beam. The implemented control systems were tested on an object, which contains the upper limb and a prototype of a loader crane, showing the greater efficiency of one control system.

Keywords: Bilateral operation · Mechatronic system · Force control

1 Introduction

The task of teleoperation systems is to increase the safety, accuracy and intuitiveness of remote machine control. The operator does not need to be at the machine to control it safely. The use of this type of control includes space, medical operations, positioning of CNC machines [1, 2]. There are the following control methods:

- Traditional method controlled by levers, joysticks, operator panels. It is the most frequently chosen control method in industry, due to the simplicity of implementation and easy access to components. Command is given by the click of a button or the manual movement of the lever [3].
- The modern type of control is voice control, the operator is able to control machine by voice itself, integrating with it two-way - he is informed about the work of the system. The principle of this system is as follows: commands are performed in the form of commands, if the voice communication operator issues the command correctly, his speech will be processed by the computer into text, and then the text will be sent to the analyser to recognize the command that was issued. Then the system needs to check if the command is safe for the environment, if so - it is executed [4].
- Gesture control is based on the machine vision, the operator, with the help of an appropriate body movement, gives commands to the controlled object. One of the interesting applications is a control LCD display during medical operations, where the doctor is analysing data during the operation (rotates the orientation of e.g. an

M. Majewski and W. Kacalak (Eds.): IIRTS 2019, LNME, pp. 171–180, 2020.
https://doi.org/10.1007/978-3-030-37566-9_16

organ virtually) - as a result, sterility is maintained in the working area (the doctor does not have to leave the table) and space is saved on the table [5].
- Haptic control uses the operator's sense of feeling. Human-machine interaction takes place in two directions - the operator controls the device cooperating with its surroundings and feels its dynamics on itself in an appropriate, safe way. An interesting application of this type of teleoperation is the repair and training of spacecraft operators who are on the planet using the above method and virtual reality. [6]. Joysticks and gaming wheels with a vibrating system are one of the simple technical solutions.

An extension of haptic methods is control using exoskeletons. In contrast to haptic joysticks, the operator performs a natural limb movement during which he is informed about the environmental resistance of a force-controlled device. The control methods presented in this paper were carried out at a specially designed and constructed test stand. The experimental stand consists of two subsystems: Master - Exoskeleton system and Slave - prototype of a loader crane.

2 Description of the System

The automatic system includes:

- executive system containing hydraulic and electric cylinders,
- real-time controller which is used for measurement processing and calculates values to control the object,
- system which has two independent objects of two articulated dynamics,
- measuring layer which is used to measure the position of joint degree by encoders, measurement of the coil current and the position of the hydraulic valve actuators.

2.1 Experiment Stand

The control system of the tested object is in a bidirectional Master-Slave configuration. The word "bidirectional" means, that the Master (system operator) controls the Slave and feels the dynamics of the Slave object and its surroundings. The Master system is an upper limb exoskeleton equipped with an electric drives (controlling by PWM signal) and a measuring system:

- positions of joint degree using incremental encoders,
- controls force measurement by strain gauges.

The construction scheme of the Master system is presented in Fig. 1. The Slave system is a prototype of the loader crane with two joints driven by hydraulic system. The loader crane is equipped with:

- executive subsystem – the hydraulic cylinders,
- control subsystem as a proportional flow valves,
- measuring subsystem including:

- the position of the hydraulic proportional valve slide using the potentiometer sensor,
- angular positions using incremental encoders,
- coil current measurement,
- measurement environmental contact point.

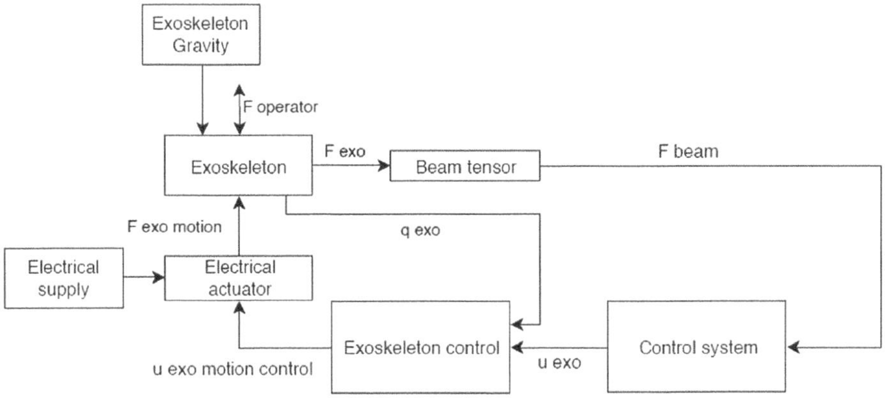

Fig. 1. Master object of the system.

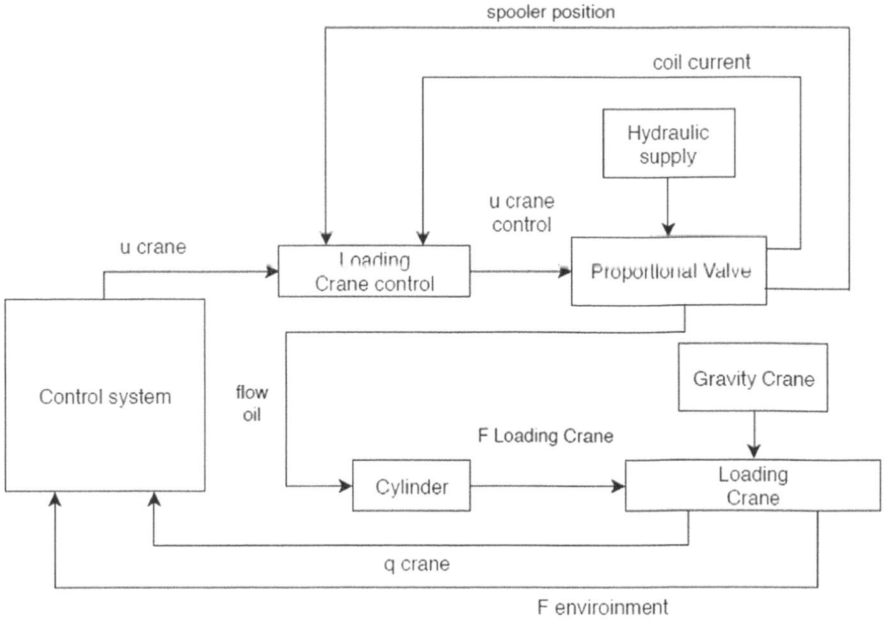

Fig. 2. Slave object of the system.

The above described subsystems are used to control the Master - Slave system. The principle of the hydraulic system is as follows (see Fig. 2): a flow oil in forced by a hydraulic pump, which is driven by electrical motor (Hydraulic supply). Input signal u_{crane} to move a loader crane comes from the Master - Slave system and next by the loader crane control block calculate $u_{crane\ control}$ signal controlling the proportional valve, which distributes oil in one of the directions. The Flow oil moves hydraulic cylinder in the proper direction. The difference pressure in a cylinder makes force $F_{Loading\ Crane}$ that moves the joints of the loader crane.

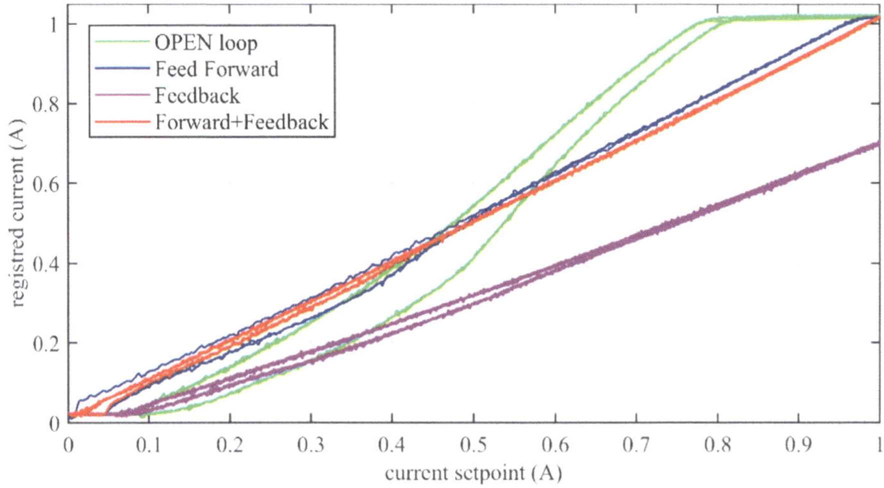

Fig. 3. Coil dynamic with different scenario control.

The system components described above are presented as real objects (Fig. 4).

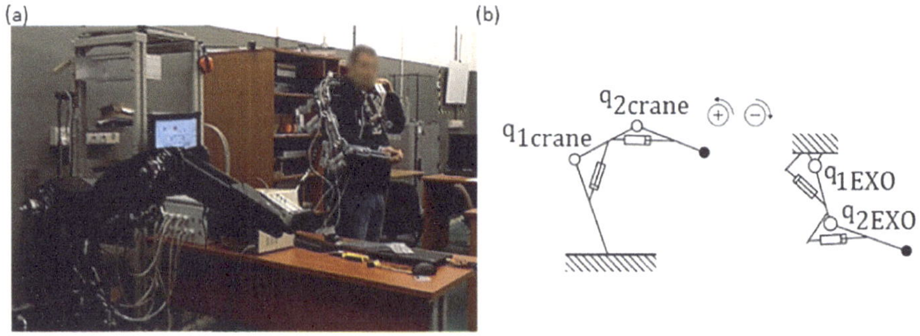

Fig. 4. (a) Experiment stand (Master, Control system and Slave), (b) Scheme of kinematic structure (Slave and Master).

The following solutions have been implemented in the automatic control system of the loader crane prototype:

- implementation of the micro-vibrations to a valve slide to eliminate static friction,
- proportional valves have a dead zone of the valve slider, in order to increase safety during manual control of the position of the valve slider by means of a lever. In the control system, this dead-zone must be removed by adding a certain constant value to the control signal,
- implementation of a static coil hysteresis loop model to improve system dynamics. The results of the implementation are described below (see Fig. 3).

Conclusion based on Fig. 3 is that the open loop system (OPEN Loop line) has a hysteresis loop. The implementation of a feedback regulator is not a satisfactory solution, increasing the regulator's settings can lead to the line approaching the 45° angle, but in reality the system becomes unstable with too much gain. Attempts have been made to implement an inverted hysteresis loop as a feed forward model, a satisfactory effect was obtained - an almost straight line was obtained at an angle of 45° angle (Feed forward line). To improve the incorrect identification of hysteresis, a coil current regulator was created, that includes the control value calculated from the feedback controller and the inverted hysteresis loop model (Feedback + Feed forward line). The feedback controller is also responsible for the dynamics at higher frequencies, because the forward model identified around the zero operating frequency (linearization point) of the valve coils.

2.2 Control Structure

By analysing the object from the point of view of the control system, the control system can be described as 3 independent elements:

- Master – an upper limb exoskeleton controlling by the system operator,
- Slave – a loader crane prototype, that performs a given trajectory of movement through the Master system, control values come indirectly from the control system,
- Control system - using the Matlab/Simulink and dSpace software, the object control values (by regulators) are calculated taking into account the safety of both machines and the operator.

The task of the control system is that the position of the exoskeleton is the same as that of the loader crane. The presented problem can be described by the equation:

$$\lim_{t \to \infty} e_q(t) = q_{crane}(t) - q_{exo}(t) = 0 \tag{1}$$

To force movement of the Slave system, it is necessary to generate tangential force acting on the strain gauges F_{beam}. Then, in the "Master - Slave control" (see Fig. 6) subsystem of the control system, values are calculated controlling the entire system:

- the u_{exo} signal generates an exoskeleton control signal informing the exoskeleton control block about the appearance of force in strain gauge F_{beam}. In the exoskeleton control block, a value is calculated to eliminate the position angle error (see Eq. 1),

- the u_{crane} signal generates a setpoint for the "loader crane control system" block in which the cascading structure of PID controllers has been implemented to ensure system stability,
- software limits ensuring system security.

3 Experiments and Results

In the experiment two control systems were tested, and the goal was to check the stability (any vibrations of the system during intermediate/transient movements) and the quality of control algorithms (see Fig. 5):

- Control algorithm without u_{exo} signal – called structure as "Feedback",
- Control algorithm with u_{exo} signal called structure as "Feedback + Feedforward",

The ITAE integral criterion of position error was used to measure the quality of the control system [7].

The experiment consisted of loading joints with a constant force in the direction of lifting and falling of the exoskeleton tip. The scheme of the experiment is shown in Fig. 5 - a red arrow indicates the direction of gravity.

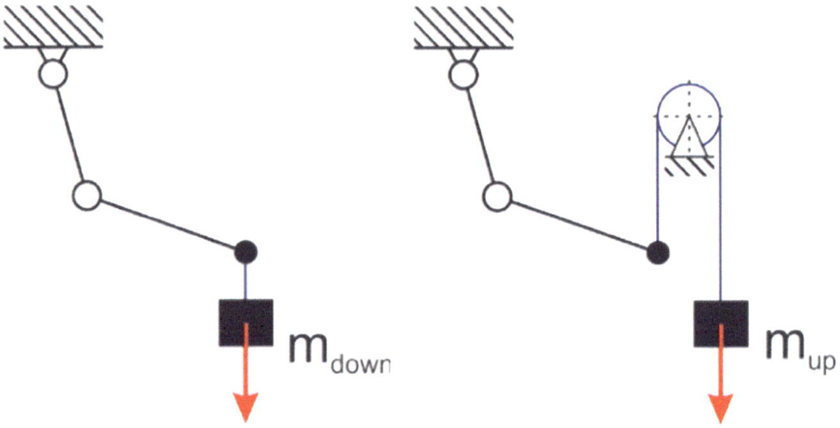

Fig. 5. Scheme of experimental stand configuration.

It was determined that the integration time of the ITAE criterion begins with forcing the force on the strain gauge beam and the final time was set to reach the extreme positions and stabilize the force measured by the strain gauge beam (any jerk in exoskeleton object).

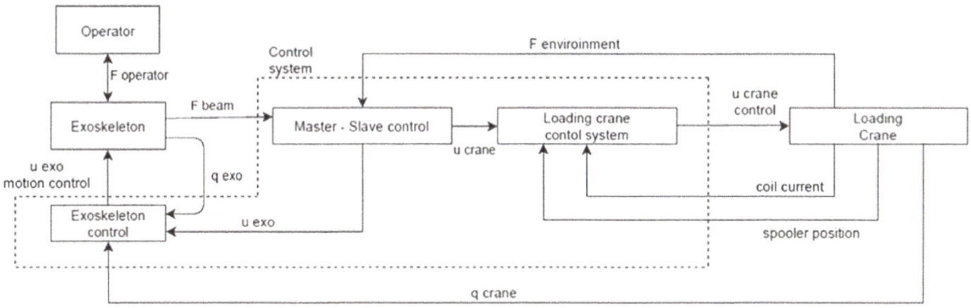

Fig. 6. Scheme of the control system.

3.1 Results of Feedback Structure Experiment

During the lifting and falling movement, the oscillation of the position of the first axis of the exoskeleton and the loader crane was noted, which can lead to a low control comfort of the system (Fig. 7).

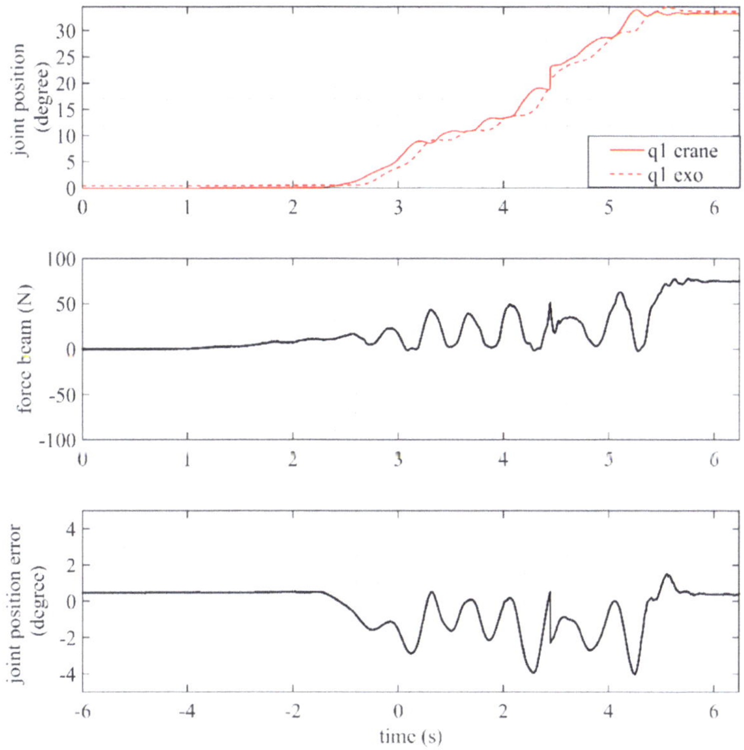

Fig. 7. Lifting tip of exoskeleton in the Feedback structure control.

The reason for these oscillations is the greater dynamics of the Master system relative to the Slave, when approaching the given position the Master executive system overtakes the Slave system by which it stops, which results in a cyclical increase in the interaction force with the operator. As a result of this nature of the work, the system is susceptible to excitation. Another reason is the high rigidity of the exoskeleton members, therefore, as it turns out later, the signal in the u_{exo} control system adds Master susceptibility, which will result in the reduced system oscillation (Fig. 8).

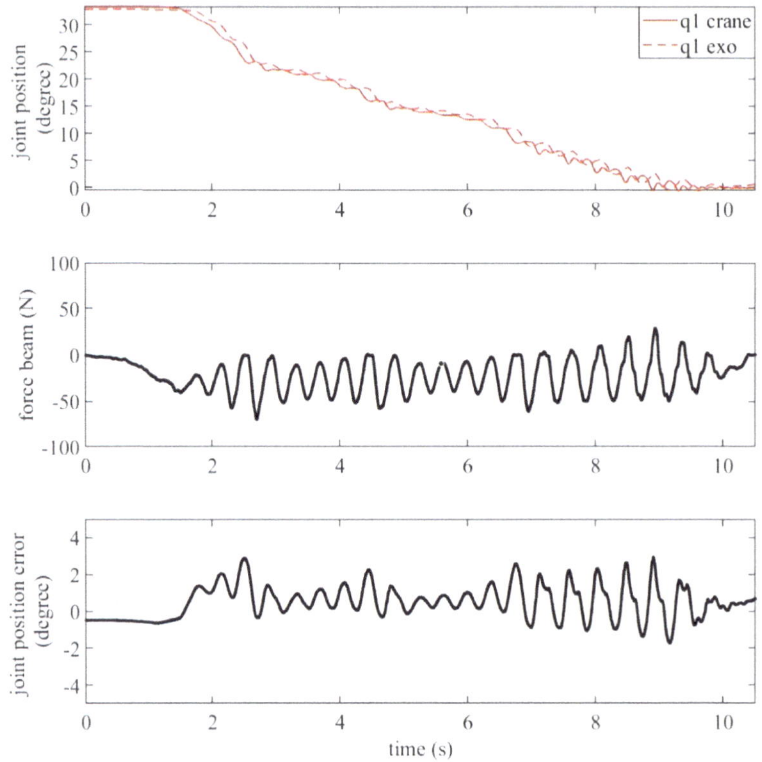

Fig. 8. Falling tip of exoskeleton in Feedback structure control.

3.2 Results of Feedback and Feedforward Structure Experiment

The automatic control system with information about the signals shows better quality of the system control (see Table 1), because the Exoskeleton control has information about the force pressed on the strain gauge beam. There was no oscillation in the system during the round trip. This type of structure is characterized by stability (no oscillations in operating states) of work. The operator is able to sense when he reaches the ends of the working space which can have a big impact on work in real conditions due to safety (Figs. 9 and 10).

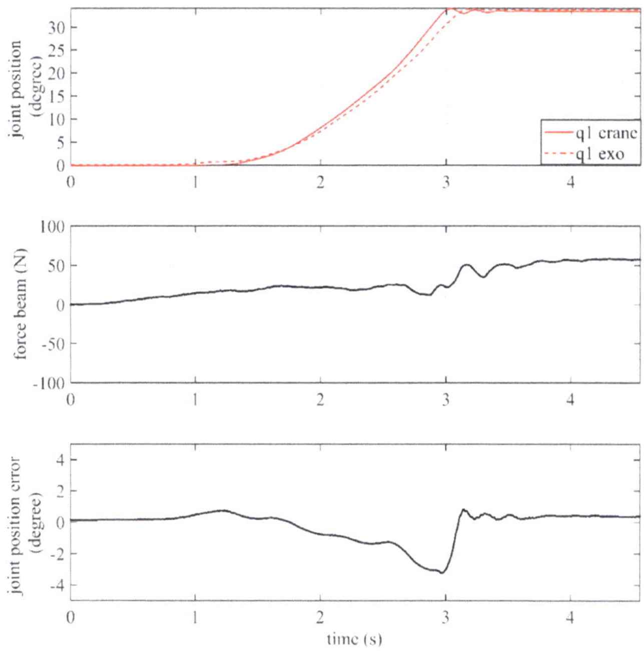

Fig. 9. Lifting tip of exoskeleton in Feedback and Feedforward structure control.

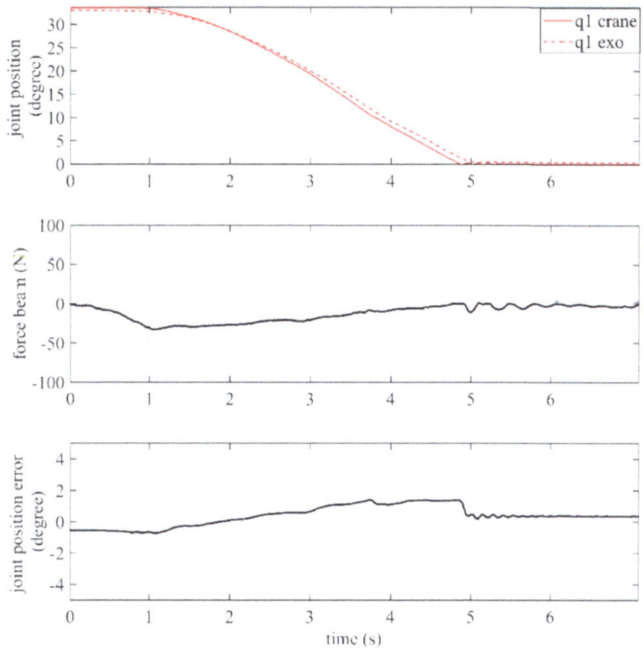

Fig. 10. Falling tip of exoskeleton in Feedback and Feedforward structure control.

Table 1. ITAE results.

	FF	FF+FB
UP	5711.60	3005.85
DOWN	8602.52	4406.95

4 Conclusions

In the presented results it can be concluded that the addition of the u_{exo} signal significantly improves the stability, quality and comfort of controlling the system. The main planned goal has been achieved. To improve the dynamics of the loader crane, proportional valves should be replaced with servo valves.

References

1. Herbin, P., Pajor, M., Stateczny, K.: Six-axis control joystick based on tensometric beam. Adv. Manuf. Sci. Technol. **40**(4), 33–41 (2016)
2. Uddin, R., Ryu, J.: Predictive control approaches for bilateral teleopration. Ann. Rev. Control **42**, 82–99 (2019)
3. Miadlicki, K., Pajor, M.: Overview of user interfaces used in load lifting devices. Int. J. Sci. Eng. Res. **6**(9), 1215–1220 (2015)
4. Majewski, M., Kacalak, W.: Intelligent speech-based interactive communication between mobile cranes and their human operators. In: Villa, A., Masulli, P., Pons Rivero, A. (eds.) Artificial Neural Networks and Machine Learning – ICANN 2016. Lecture Notes in Computer Science, vol. 9887. pp. 523–530. Springer, Cham (2016)
5. Huber, T., Wunderling, T., Paschold, M., et al.: Highly immersive virtual reality laparoscopy simulation: development and future aspects. Int. J. Comput. Assist. Radiol. Surg. **13**(2), 281–290 (2018)
6. Penn, P., Petrie, H., Colwell, Ch., Kornbrot, D., Furner, S., Hardwick, A.: The haptic perception of texture in virtual environments: an investigation with two devices. In: Proceedings of the First International Workshop on Haptic Human-Computer Interaction, pp. 25–30 (2000)
7. Martins, F.G.: Tuning PID controllers using the ITAE criterion. Int. J. Eng. Educ. **21**(5), 867–873 (2005)

Teaching Study of Engineering Graphics for Expressions of Innovative Design

Wanghui Bu(✉)

School of Mechanical Engineering, Tongji University, Shanghai 201804, China
buwanghui@tongji.edu.cn

Abstract. Graphics are used early to express design ideas. Early engineers were often painters, and their drawings pursued vivid images. However, such a drawing method requires a lot of time and effort, and this graphics technique is difficult to grasp by ordinary engineers. Until the "Descriptive Geometry" by Monge came out, the method of expressing the spatial object by orthogonal projection on the plane was adopted by the majority of engineers. With the development of computer technology, engineering graphics now have more expression approaches and can more effectively express innovative design ideas. The engineering graphics course should develop students' capabilities of spatial imagination, machine expression, drawing software application, and preliminary innovative design. The cultivation of the first three abilities is relatively easy to achieve since these are the traditional teaching goal of the engineering graphics course, but the cultivation of innovative design ability is relatively difficult. This paper investigates the development of graphics technologies, classifies the graphics technologies according to the information dimensions, and studies the effect of different graphic technologies on the expression of innovative design. From the perspective of student learning, three transitions of the learning method to become an excellent engineer are required to achieve: from passive learning to active learning, from subject learning to project learning, from definitive learning to tentative learning. The teaching practice shows that, the innovative design based on graphic technology is an effective way to realize the transitions of the learning method, and the graphics technology has significantly cultivated students' innovative design ability.

Keywords: Engineering graphics · Innovative design · Teaching and learning

1 Introduction

Training excellent engineers is an important goal of higher education in mechanical engineering. As a professional basic course for college students majoring in mechanical engineering, the course of engineering graphics has both theory and practice, and it plays an important role in the construction of the profession. The teaching quality of engineering graphics is related to the study of students' follow-up professional courses and the cultivation of professional qualities.

The engineering graphics course should develop students' capabilities of spatial imagination, machine expression, drawing software application, and preliminary innovative design. The cultivation of the first three abilities is relatively easy to

M. Majewski and W. Kacalak (Eds.): IIRTS 2019, LNME, pp. 181–190, 2020.
https://doi.org/10.1007/978-3-030-37566-9_17

achieve, since this is the traditional teaching goal of engineering graphics. However, the cultivation of innovative design abilities is relatively difficult, especially for the college students in lower grades. Many excellent engineering graphics textbooks hope to cultivate students' innovative design capabilities. "Technical Drawing with Engineering Graphics" by Giesecke et al. [1] not only explains the various 2D and 3D expression techniques in engineering graphics, but also introduces the patent drawings and the design process of mechanical products. In order to cultivate students' design ability, "Machine Drawing" by Narayana et al. [2] successively teaches blueprint reading, assembly drawing, and finally part drawing. "Manual of Engineering Drawing" by Simmons et al. [3] introduces the latest technology in engineering graphics, including 3D annotation of part dimensions and technical requirements. "Mechanical Drawing" by He et al. [4] cultivate students' imagination and creativity through conception of 3D shapes.

Although a few excellent engineering graphics textbooks involve some training of innovative design ability, most engineering graphics teaching mainly teaches reading and drawing skills from the perspective of manufacturing, and insufficient training for students' innovative design. This paper investigates the development of graphics technology, classifies the graphics technologies according to the information dimensions, and studies the effect of different graphics technologies on the expression of innovative design. From the perspective of student learning, this paper presents that three types of transitions of learning methods are required for the student to become an excellent engineer. Through teaching practice, the innovative design based on graphics technology is an effective way to realize the transitions of learning methods.

2 Development of Engineering Graphics Technology

People used graphics early to express design ideas. Early expressions have no specific drawing rules and are entirely dependent on the designer's drawing. In order to effectively express design ideas, early engineers were often painters, and their drawings pursued vivid images and strong stereoscopic effects. Such a drawing method requires a lot of time and effort, and this drawing technique is difficult to grasp by ordinary engineers. Until the advent of "Descriptive Geometry", the method of expressing the spatial object by orthogonal projection on the plane was adopted by majority of engineers. With the development of computer technology, engineering graphics now have more expression methods and can more effectively express innovative design ideas.

2.1 Leonardo da Vinci's Era

Before the advent of "Descriptive Geometry", in order to effectively express design ideas, the designers' drawings were realistic and vivid. Leonardo da Vinci (Italian, 1452–1519) 500 years ago is their outstanding representative. He has designed many mechanical devices. Figure 1(a) is his work "Mechanical Wings" designed in 1494, and Fig. 1(b) is a three-dimensional modeling of Leonardo's "Mechanical Wings" by Italian researchers using modern design methods [5]. It can be seen that the painters in the Leonardo's era were not only artists, but many of their paintings had high engineering practical value.

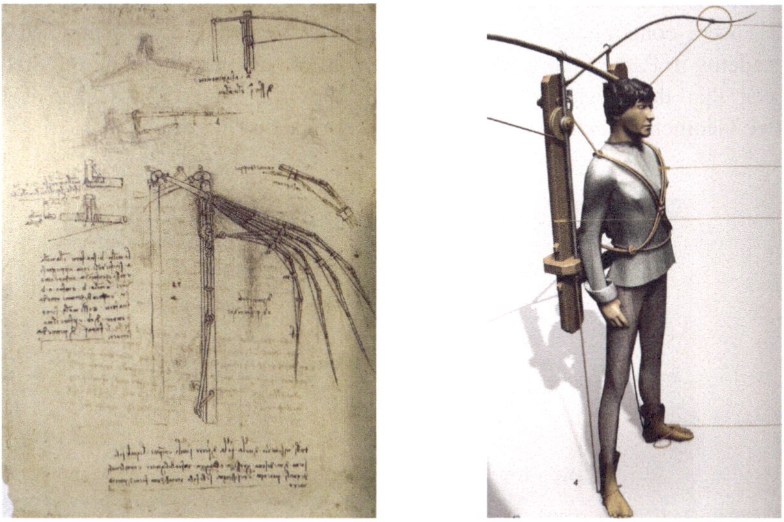

(a) Leonardo's "Mechanical Wings" (b) 3D modeling of "Mechanical Wings"

Fig. 1. Leonardo da Vinci's work "Mechanical Wings" drawn in 1494.

There is a question worth considering. Although many of the mechanical devices designed by Leonardo da Vinci had great value at the time, in terms of "mechanical wings", Leonardo hoped that people could wear the wings and were able to be like birds in the sky. Flying, this was not possible not only 500 years ago, but even today it is difficult to achieve. Therefore, can this kind of unrealized work count as an innovative design? We believe that as long as the work expresses design details, has design basis, and is innovative in function and structure, it can be counted as an innovative design. Our judgment on innovative design works mainly considers the following two factors. First, today's innovative design products require multiple design iterations, and it is difficult to ensure that the original design can make mature products. Second, in terms of technological advancement, based on the mechanism and structure of "Mechanical Wings" designed by Leonardo, the addition of small-volume yet high-torque motors and intelligent control systems may make the "Mechanical Wings" be realized in the near future.

2.2 Napoléon-Monge's Era

Drawing techniques like Leonardo da Vinci require a lot of time and effort and are difficult to master by ordinary engineers. Until more than 200 years ago, the "Descriptive Geometry" written by Monge (French, 1746–1818) came out, and the method of expressing the spatial object by orthogonal projection on the plane was adopted by majority of engineers. Today's engineering graphics textbooks still use this accurate, clear, and measurable two-dimensional drawing method. Monge's "Descriptive Geometry" was originally taught in French military academies in 1769. It was a military secret and was not published until 1798. Monge and Napoléon (French, 1769–1821) were contemporaries. Napoléon learned "Descriptive Geometry" at the military academy in Paris. Figure 2 shows the design of Napoléon's throne in 1804. It can be seen that this design drawing adopted the front and right views of the "Descriptive Geometry" technique to express design idea of the throne.

Fig. 2. The design drawing of Napoléon's throne in 1804 (using front and right views to express design ideas).

2.3 Graphics Technologies According to Information Dimensions

According to the information dimensions contained in the drawing, the graphics technologies can be classified as the following categories:

(1) 2D views and sectional views. Regardless of computer 2D drawing, ruler drawing, or freehand drawing, the most common drawings in engineering are presented by the 2D views and the sectional views.
(2) 2.5D line stereo views. Draw a stereo view of an object on a plane with lines or curves. This kind of view seems to have a strong three-dimensional sense, such as the isometric or perspective view, but is still drawn in a two-dimensional plane. We refer to these isometric or perspective views as the 2.5D line stereo view.
(3) 3D solid views. The 3D solid view is obtained by creating a solid model in the three-dimensional graphics space by the computer and then projecting it to a plane. It includes 3D part view and 3D assembly view.
(4) 4D motion view. The information dimensions of the above three categories are all lengths. If the time dimension is added, the motion view of a mechanical device changing with time can be obtained.
(5) 5D motion color view. On the basis of the 4D motion view, other information such as stress or strain can be added into the view by color. In this way, the motion color view of a mechanical device is obtained.

It is noted that the above graphics technologies are classified as five categories according to the information dimensions, but it does not mean that the low-dimensional graphics expression method is not important. In fact, 2D and 2.5D graphics technologies are still very important in the early stage of innovative design, especially for the rapid expression of creative ideas. Figure 3(a) and (b) are respectively the sectional view of the tableware supply device and the perspective view of the vending device for an unmanned canteen designed by students.

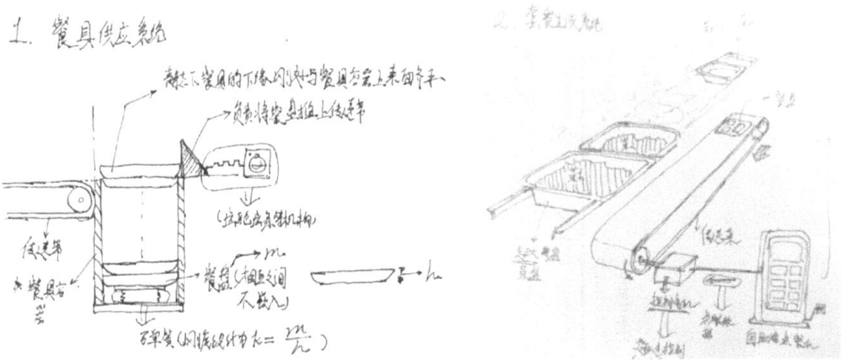

(a) sectional view (b) perspective view

Fig. 3. The sectional view of the tableware supply device and the perspective view of the vending device for an unmanned canteen.

Figure 4(a) and (b) are the 3D solid views of the rear wheel steering of a four-wheel steering bus with mechanical transmission designed by students, which are also two segments of the 4D motion view of the wheel steering process. The four-wheel steering bus realizes the compatible steering of four wheels through mechanical transmission. The advantage is that the turning radius is small and the turning of the bus is more flexible.

(a) a segment of the rear wheel steering (b) another segment of the rear wheel steering

Fig. 4. 3D solid views of the rear wheel steering of a four-wheel steering bus with mechanical transmission.

3 Transition of Learning Methods for Cultivation of Innovative Design Capabilities

The aim of cultivate students in mechanical engineering is to enable them to become excellent engineers. Here we classify learning methods according to three dimensions, which are passive learning and active learning, subject learning and project learning, and definitive learning and tentative learning, as shown in Fig. 5 [6].

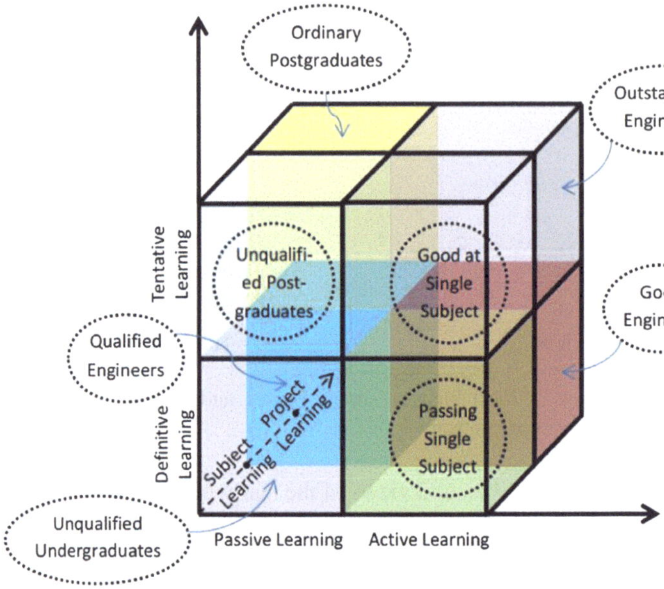

Fig. 5. The three-dimensional model of learning methods.

3.1 Passive Learning and Active Learning

Passive learning means that students are guided step by step by the teacher, and the subjective initiative of students is poor. Active learning means that students are not only satisfied with the teacher's classroom teaching but also actively read the materials and hands-on practice, and students' subjective initiative is very strong. For the transition of the learning methods, the transition from passive learning to active learning is the most basic change in learning methods for training to become an outstanding engineer.

3.2 Subject Learning and Project Learning

Subject learning means that students follow the curriculum and learn each course to obtain credits. Project learning refers to students' comprehensive application of various course knowledge and completion of engineering projects or research projects. For the transition of the learning methods, the transition from subject learning to project learning is the most important change in learning methods for cultivating to become an outstanding engineer.

3.3 Definitive Learning and Tentative Learning

Definitive learning refers to students, especially undergraduates, learning books or the knowledge taught by teachers. Tentative learning means that students, especially postgraduates, learn about knowledge beyond the classroom knowledge through exploratory experiments. For the transition of the learning methods, the transition from definitive learning to tentative learning is the most difficult change in learning methods for training to become an excellent engineer.

4 Innovative Design Based on Graphics Technology

In order to cultivate excellent engineers, at the teaching level, we use graphic technology as a tool to build a transition path from passive learning to active learning, from subject learning to project learning, from definitive learning to tentative learning. Graphic technology can be associated with multiple courses. Teaching practice shows that graphic technology has significantly cultivated students' innovative design capabilities, and many students have won awards in various innovative design competitions.

4.1 Advantages of Graphics Technology

Innovative design based on graphics technology refers to the use of 2D or 2.5D (such as freehand sketch, drawing by rulers, AutoCAD computer drawing), 3D (computer modeling software, such as SolidWorks, Creo, Inventor, CATIA, NX), 4D and 5D (such as the motion simulation module of the 3D software, professional animation software CINEMA 4D) graphics technologies for the 2D drawing, 3D modeling, and motion simulation of mechanical devices. Advantages of innovative design based on graphics technology include:

(1) Intuitive expression. Especially the three-dimensional models and four-dimensional motion simulation animations are very vivid and can stimulate the interest of students.
(2) Convenient modification. The digital design by the computer can be easily modified without being reworked like physical production, which is suitable for beginners in engineering education.
(3) Low cost. Students only need to use drawings or computers. There are no manufacturing costs. Therefore, this is suitable for undergraduate junior students who lack financial support.
(4) Easy for accumulation of experience. Traditionally, the accumulation of experience for mechanical engineers requires a lot of time, since the cycle of product design, manufacturing, assembly, and commissioning is very long. The innovative design based on graphic technology can focus on the innovative ability cultivation in the

design stage, and verify the feasibility of the innovative design through computer simulation technology, so the time required to accumulate experience is greatly reduced.

Therefore, the innovative design based on graphic technology is very suitable for the undergraduate students to change from passive learning to active learning. If combined with the relevant digital or mechanical innovative design competitions, the learning method of students can be changed from subject learning to project learning. Some excellent students can even realize the transition from definitive learning to tentative learning.

4.2 Teaching Program of Engineering Graphics

In order to enhance the important role of engineering graphics in the expression of innovative design, we have explored the longitudinal extension and the horizontal integration of engineering graphics course, as shown in Fig. 6.

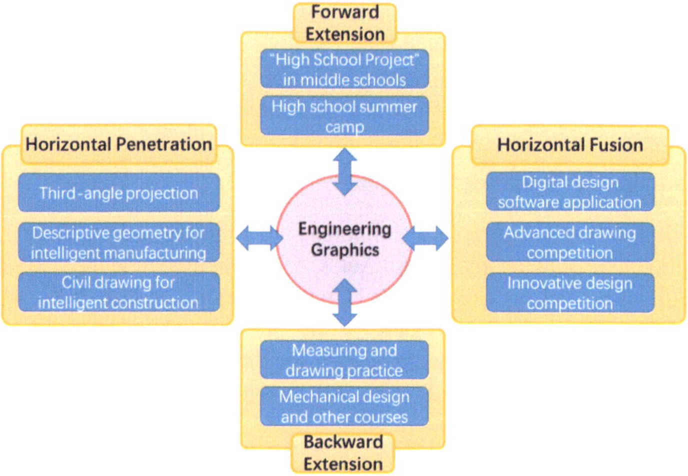

Fig. 6. Longitudinal extension and horizontal integration of engineering graphics.

The longitudinal extension of the engineering graphics course is divided into forward extension and backward extension. On the one hand, the forward extension refers to the "High School Project" established in the relevant middle schools. It selects excellent professors to give lectures in middle schools, and teaches high school students the basic knowledge and typical applications of engineering graphics. On the other hand, the forward extension refers to the selection of high school students during the summer. The selected students participate in the summer camp of our college. In the summer camp, excellent lecturers teach high school students the theoretical basis of engineering graphics and the application skills of computer graphics. The backward extension refers to the use of engineering graphics technology in subsequent professional courses, including measuring and drawing practice, mechanical design and other courses.

The horizontal integration of engineering graphics is divided into horizontal penetration of theoretical knowledge and horizontal fusion of application practice. The horizontal penetration of theoretical knowledge refers to the teaching of the third-angle projection method adopted by the United States and Japan, in addition to the first-angle projection method adopted by China. Besides teaching classical engineering graphics, it also includes the descriptive geometry for intelligent manufacturing. The horizontal fusion of application practice is to guide students to apply digital design software to participate in the advanced engineering graphics competitions and mechanical innovative design competitions after class.

4.3 Students' Innovative Design Results

The teaching practice of innovative design based on graphics technology shows that it has significantly cultivated students' innovative design ability, and many students have won awards in various innovative design competitions. As a representative work thereof, Fig. 7(a) shows a 3D solid view of a two-layer rotary bicycle library designed by a student, and Fig. 7(b) is a prototype model manufactured. The upper and lower layers can be rotated by themselves to facilitate parking and picking up the bicycle; the upper layer is supported by a gas spring for easy pull-down; the entire device is purely mechanical and requires no external power. This work won the first prize in the mechanical innovative design competition.

(a) 3D solid view of the bicycle library (b) prototype model of the bicycle library

Fig. 7. The two-layer rotary bicycle library.

5 Conclusions

Since "Descriptive Geometry" by Monge came out, the method of expressing the spatial object through the orthogonal projection on the plane has been adopted by the majority of engineers. This paper investigates the development of graphics technologies, classifies the graphics technologies according to information dimensions, and studies the effect of different graphics technologies on the expression of innovative design. With the development of computer technology, engineering graphics now have more expression approaches and can more effectively express innovative design ideas.

From the perspective of student learning, the engineering graphics course should develop students' capabilities of spatial imagination, machine expression, drawing software application, and preliminary innovative design. The cultivation of the first three abilities is relatively easy to achieve, but the cultivation of innovative design ability is relatively difficult. This paper proposes three transitions of the learning method to become an excellent engineer: from passive learning to active learning, from subject learning to project learning, from definitive learning to tentative learning. Teaching practice shows that innovative design based on graphics technology is an effective way to realize the transitions of the learning method, and the graphics technology has significantly cultivated students' innovative design ability.

Acknowledgements. This study is supported by Postgraduate Education Reform Projects of Tongji University (Grant Nos. 2018GH030, ZD190403, ZD19020103), Teaching Reform Research and Construction Project of Tongji University, and Fundamental Research Funds for the Central Universities.

References

1. Giesecke, F., Mitchell, A., Spencer, H., et al.: Technical Drawing with Engineering Graphics, 15th edn. Peachpit Press, USA (2016)
2. Narayana, K., Kannaiah, P., Venkata Reddy, K.: Machine Drawing, 3rd edn. New Age International, India (2009)
3. Simmons, C., Maguire, D., Phelps, N.: Manual of Engineering Drawing, 3rd edn. Elsevier, UK (2009)
4. He, M., Qian, K., Xu, Z.: Mechanical Drawing, 7th edn. Higher Education Press, China (2016)
5. Laurenza, D.: Leonardo's Machines: Secrets and Inventions in the Da Vinci Codices. Giunti Editore, Italy (2005)
6. Bu, W.: Innovation-oriented transition of learning methods for undergraduates and postgraduates in mechanical engineering. In: 7th International Research Symposium on PBL, pp. 48–57. Aalborg University Press, Denmark (2018)

Author Index

M. Majewski and W. Kacalak (Eds.): IIRTS 2019, LNME, p. 191, 2020.
https://doi.org/10.1007/978-3-030-37566-9

The manufacturer's authorised representative in the EU is Springer
Nature Customer Service Centre GmbH, Europaplatz 3, 69115 Heidelberg,
Germany. If you have any concerns regarding our products, please
contact ProductSafety@springernature.com

Printed and bound by CPI Group (UK) Ltd, Croydon, CR0 4YY
30/09/2026
02225448-0004